Selected Topics on Computer-Aided Process Design and Analysis

R. S. H. Mah and G. V. Reklaitis, editors

Gary Agin
T. Bailey
T. E. Baker
Gary L. Blau
N. L. Book
D. R. Evans
L. A. Fabiano
L. W. Fish
P. Gacka
J. L. Gaddy
C. M. Hilton
R. A. Knudsen
J. M. Langa

W. W. Madsen
D. L. Martin
M. J. Kolbet Mattione
W. J. Weier
P. A. Minderman, Jr.
J. M. Neville
J. D. Perkins
G. V. Reklaitis
R. W. H. Sargent
V. B. Shah
M. A. Stadherr
D. W. Tedder
R. Ventker

AIChE Symposium Series

Number 214   1982   Volume 78

Published by

American Institute of Chemical Engineers

345 East 47 Street   New York, New York 10017

Copyright 1982

American Institute of Chemical Engineers
345 East 47 Street, New York, N.Y. 10017

Library of Congress Cataloging in Publication Data

Main entry under title:
   Selected topics on computer-aided process design and analysis.

   (AIChE symposium series; no. 214)
   Contents: Speedup: a computer program for steady-state and dynamic simulation and design of chemical processes/J. D. Perkins and R.W.H. Sargent — Development of a new equation-based process flowsheeting system: numerical studies/M.A. Stadtherr and C.M. Hilton — An equation-oriented approach to the structuring and solution of chemical process design problems/M.J. Kolbet Mattione, W.J. Meier, and N.L. Book — [etc.]
   1. Chemical processes—Data processing—Congresses.
   I. Mah, Richard S. H. II. Reklaitis, G. V., 1942- III. Series.
TP155.7.S425 1982   660.2'81'02854   82-11352
ISBN 0-8169-0225-9   ISSN 0065-8812

The appearance of the code at the bottom of the first page of an article in this serial indicates the copyright owner's consent that for a stated fee copies of the article may be made for personal or internal use or for the personal or internal use of specific clients. This consent is given on the condition that the copier pay the per-copy fee (appearing as part of the code) through the Copyright Clearance Center, Inc., 21 Congress St., Salem, Mass. 01970, for copying beyond that permitted by Sections 107 or 108 of the U.S. Copyright Law. This consent does not extend to copying for general distribution, for advertising or promotional purposes, for inclusion in a publication, or for resale.

Articles published before 1978 are subject to the same copyright conditions, except that the fee is $2.00 for each article.

Printed in the United States of America by
Lew A. Cummings Co., Inc.

FOREWORD

This volume contains papers selected from six of the sessions sponsored by the CAST (Computing and Systems Technology) Division at the New Orleans (November 1981) and Orlando (March 1982) AIChE meetings. For the past several years CAST Division programming activities have grown significantly. The number of papers presented at meetings far outstrips the capacity of channels for publications. At the same time escalating travel expenses prevent even broader member participation at these meetings. Papers which are oriented towards applications are frequently unavailable in printed form, and yet often are of considerable interest to industrial practitioners in our field. This volume is an experiment undertaken to respond to the need expressed by many industrial members to have timely dissemination of such technical information in printed form.

The topics covered in this volume ranged from research on computer-aided process design systems to application experience with existing tools. The first three papers address the numerical and computational aspects of equation-based process flowsheeting systems. They are followed by three papers reporting on the experience of applying ASPEN, a recently developed sequential modular process simulator. The next two papers deal with optimal design and optimization techniques. The discussion of the design and analysis of continuously operated processes is rounded out by the paper on the use of a continuous simulation language to reactor design. The last two papers deal with the scheduling of batch and semi-continuous plants, which are of growing interest to chemical engineers. The diversity of these topics is a reflection of the many parallel developments which are rapidly taking place in our field.

We wish to acknowledge the assistance of the reviewers of individual papers and of Professor J. D. Seader and Dr. J. J. Siirola who, as chairman of their respective sessions, guided us in our selection of material. We alone bear the ultimate responsibility for the final selection.

R.S.H. Mah
Department of Chemical Engineering
Northwestern University
Evanston, Illinois

G.V. Reklaitis
Department of Chemical Engineering
Purdue University
West Lafayette, Indiana

CONTENTS

FOREWORD .... *iii*

SPEEDUP: A COMPUTER PROGRAM FOR STEADY-STATE AND DYNAMIC SIMULATION AND DESIGN OF CHEMICAL PROCESSES ....
.... J. D. Perkins and R. W. H. Sargent   1

DEVELOPMENT OF A NEW EQUATION-BASED PROCESS FLOWSHEETING SYSTEM: NUMERICAL STUDIES .... M. A. Stadtherr and C. M. Hilton   12

AN EQUATION-ORIENTED APPROACH TO THE STRUCTURING AND SOLUTION OF CHEMICAL PROCESS DESIGN PROBLEMS ....
.... M. J. Kolbet Mattione, W. J. Meier and N. L. Book   29

EXPERIENCE WITH ASPEN WHILE SIMULATING A NEW METHANOL PLANT ....
.... R. A. Knudsen, T. Bailey and L. A. Fabiano   38

THE APPLICATION OF ASPEN FLOWSHEET SIMULATOR AT ALCOA ....
.... V. B. Shah, P. Gacka and J. M. Langa   56

THE USE OF ASPEN IN THE ANALYSIS OF THERMODYNAMIC CYCLES ....
.... L. W. Fish, D. R. Evans and W. W. Madsen   66

COMPARISONS OF DISTILLATION NETWORKS — EXTENSIVELY STATE OPTIMIZED VERSUS EXTENSIVELY ENERGY INTEGRATED ....
.... P. A. Minderman, Jr. and D. W. Tedder   69

THE OPTIMAL DESIGN OF RESILIENT HEAT EXCHANGER NETWORKS ....
.... A. R. Parkinson, J. S. Liebman, C. O. Pedersen and A. B. Templeman   85

PROCESS OPTIMIZATION WITH THE ADAPTIVE RANDOMLY DIRECTED SEARCH
.... D. L. Martin and J. L. Gaddy   99

APPLICATION OF DACSL (DOW ADVANCED CONTINUOUS SIMULATION LANGUAGE) TO THE DESIGN AND ANALYSIS OF CHEMICAL REACTOR SYSTEMS ..
.... Gary Agin and Gary L. Blau   108

REVIEW OF SCHEDULING OF PROCESS OPERATIONS .... G. V. Reklaitis   119

PROSIT — AN INTERACTIVE PROCESS SCHEDULING SYSTEM ....
.... J. M. Neville, R. Ventker and T. E. Baker   134

SPEEDUP: A COMPUTER PROGRAM FOR STEADY-STATE AND DYNAMIC SIMULATION AND DESIGN OF CHEMICAL PROCESSES

J.D. PERKINS
and
R.W.H. SARGENT

Imperial College
London SW7 2BY
England

Abstract

The paper describes the development of a computer program which permits steady-state and dynamic simulation and optimization of chemical processes. The program allows the user a great deal of flexibility in the way he defines a model for his process, since it will deal with an arbitrary mixture of equations and procedures (subroutines) relating variables in the process.

The first part of the paper gives a brief description of the facilities available in SPEEDUP, and of the language the user employs to define his problem. This description is followed by a report on some of the recent research done at Imperial College on efficient and robust numerical methods to handle the solution of these problems. Results of some tests of new quasi-Newton methods for sets of nonlinear equations are presented, and comparative tests on methods for solving sparse problems are reported. The development of software for mixed systems of differential and algebraic equations is outlined.

1. Introduction

Process "flowsheeting" packages have long since become a standard tool for steady-state process design, but with the growth of interest in control system design, hazard analysis, and operability studies, there has developed a parallel need for dynamic simulation packages. The latter have evolved quite independently of the activity in steady-state flowsheeting, having their roots in digital simulators of analogue computers, so that model forms and data requirements are quite different for the two systems.

Clearly this is a gross inconvenience for the user, since much of the basic information is required for both, in spite of differences in use and differences in level of sophistication of the models commonly employed. Thus at Imperial College in 1979 we launched a programme to develop a new package which can be used for both steady-state simulation or process design and dynamic simulation or control system design. The programming effort itself is backed up by a series of researches on the underlying numerical techniques, and this paper is a preliminary report on the whole project.

In the context of steady-state design, much has been written on the relative merits of procedure-oriented and equation-oriented systems, and the arguments on both sides have been summarized by Hernandez and Sargent (1979). The same dichotomy exists in dynamic simulation systems, with protagonists for both approaches. It seems sensible to provide for both, and indeed allow the user an arbitrary mixture of procedures and equations in describing his system. This has implications at all levels - the input language, the systems analysis, and the numerical techniques - in order to avoid serious loss of efficiency.

There is also the question of whether one opts for a general-purpose simulation system or tailors the structure to the specific needs of chemical process systems. Again we have tried to gain the best of both worlds with a two-level approach - a flexible general-purpose facility at the systems level, with a data-structure and input language which provides the natural context for the process engineer as user. The system is of course interactive, though provision is made for running portions in batch mode. It is also modular, so that the user can change not only data and operating conditions, but also the level of sophistication of the models for the various parts of the plant in the light of the results.

2. A Brief Description of SPEEDUP

The core of the SPEEDUP system is a data-base representing the current state of the design. Input information and commands for carrying out the various phases of the design are expressed in a special-engineering-oriented language. The system interfaces with a data-bank which contains design data (plant unit models, physical properties, standards, costs) and a library of FORTRAN subroutines (physical property correlations, numerical procedures) used by the executive. Facilities are provided to enable the user to add to or modify this data-bank.

The information representing the current state of the design is of several distinct kinds, each of which is stored in tis own section of the data-base. The sections we shall consider here are those to do with specifying the simulation problem:

FLOWSHEET   The flow-diagram of the process, indicating the connections between the various units.

MODEL   The describing equations for each of the unit-types occurring in the process.

UNIT   The design specifications for each unit or process stream.

OPERATION   The operating policy to be followed during the simulation.

A brief description of some of the statements used to define the information is given below. The statements are illustrated by reference to a specific example, the flow-diagram for which is shown in Figure 1. The pressure of a stream containing a mixture of hydrocarbons is reduced in two stages. As the pressure is reduced, some of the more volatile components flash off. We shall consider the problem of determining first a steady-state of the process and then what happens if the feed pressure varies. The FLOWSHEET, MODEL and UNIT sections of the data-base are the same for these two problems.

2.1 FLOWSHEET

The first step in the specification of problems to SPEEDUP is to define the process flow-diagram. Each unit in the flowsheet is named, and the connections between units are specified using link statements, for example:

OUTPUT OF V1 IS INPUT OF D1

Where a unit has several inputs and/or outputs, these may be labelled by numbering as in

OUTPUT 2 OF D1 IS INPUT OF V2

As well as inputs and outputs, units may have CONNECTIONS. This term can be used to differentiate between streams transferring material, and other connections transferring information (e.g. control lines).

The full set of statements defining the flowsheet in Figure 1 is shown in Figure 2.

2.2 MODEL

Each unit-type has a MODEL section which gives the describing equations relating the input and output stream variables and the unit design parameters. An example of a MODEL section, specifying the flash drums in our problem, is shown in Figure 3. All text enclosed by the symbols #...# is treated as commentary by SPEEDUP. As with a FORTRAN subroutine, the variables used in defining the model are specific to the model, and can be simple variables or subscripted variables (members of arrays). As well as the describing equations themselves, we must also define the variables associated with each input and output stream, and as in FORTRAN we must define the dimensions of any arrays of variables so that the appropriate storage can be allocated. The MODEL section therefore has various subsections:

SET

This section is used to declare those variables (e.g. array dimensions) which must be set in order for the model to be defined.

In addition the SET section may be used to give values to physical constants appearing in the model.

ARRAY

In this section, the arrays used in the model are declared and dimensioned. Note that we may use parameters declared in the SET section as dimensions.

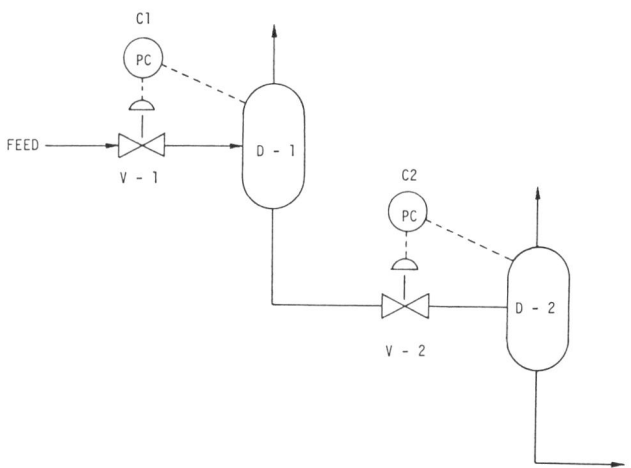

Figure 1. Flowsheet of a pressure reduction system.

FLOWSHEET

```
OUTPUT OF FEEDER IS INPUT OF V1

OUTPUT OF V1 IS INPUT OF D1

OUTPUT 1 OF D1 IS PRODUCT 1

OUTPUT 2 OF D1 IS INPUT OF V2

OUTPUT OF V2 IS INPUT OF D2

OUTPUT 1 OF D2 IS PRODUCT 2

OUTPUT 2 OF D2 IS PRODUCT 3

CONNECTION OF V1 IS OUTPUT OF C1

CONNECTION 1 OF D1 IS INPUT OF C1

CONNECTION OF V2 IS OUTPUT OF C2

CONNECTION OF D2 IS INPUT OF C2
```

Figure 2. An example of a FLOWSHEET section.

```
MODEL FLASHDRUM
   # DYNAMIC BEHAVIOUR OF A FLASH DRUM #
   # EQUILIBRIUM ASSUMED BETWEEN VAPOUR AND LIQUID OUTPUTS #
SET N, G = 9.81
ARRAY BOT(N), TOP(N), X(N), Y(N), FEED(N), HOLDUP(N)
STREAMS
      INPUT      1 FEED, ENTHF, P
      OUTPUT     1 TOP,  ENTHV, PTOP
      OUTPUT     2 BOT,  ENTHL, PBOT
      CONNECTION 1      P
      CONNECTION 2      TEMP
      CONNECTION 3      L
EQUATION
      FOR ALL I TO N
   # BOTTOMS COMPONENT FLOWS IN TERMS OF COMPOSITION #
   # OF LIQUID HOLDUP #
      BOT(I) = X(I)*BOTT
   # TOPS FLOWS IN TERMS OF VAPOUR COMPOSITION #
      TOP(I) = Y(I)*TOPT
   # DYNAMIC BALANCE FOR TOTAL HOLDUP OF EACH COMPONENT #
      £HOLDUP(I) = FEED(I) - TOP(I) - BOT(I)
   # DYNAMIC ENTHALPY BALANCE #
      £ENTH = ENTHF * SIGMA(FEED) - ENTHV*TOPT - ENTHL*BOTT
   # TOPS AND BOTTOMS FLOWS FROM PRESSURE DROPS #
      TOPT = RHOV*CV*SQRT(P-PTOP)
      BOTT = RHOL*CV*SQRT(P+L*G/AREA/RHOL/1.E5
             *DENS(X) - PBOT)
   # PROCEDURE TO DETERMINE EQUILIBRIUM COMPOSITION OF #
   # VAPOUR AND LIQUID #
PROCEDURE
      (P,TEMP,V,L,ENTHV,ENTHL,RHOV,RHOL,X,Y) FLASHV
                     (VOLUME,ENTH,HOLDUP)
```

Figure 3. An example of a MODEL section.

STREAMS

Here the variables associated with inputs,

outputs and connections to the unit are declared. Note that input and output streams do not have a fixed format, but can be any collection of variables in the model. Thus a 'model' need not represent a unit of plant, but is simply a collection of equations relating sets of variables. Furthermore, since the model is not a procedure, there is no directionality implied by the definitions of 'output' streams. Therefore SPEEDUP can be used to handle any type of network, or interrelated system of equations. The nomenclature employed is simply to conform with the usual way of describing process flowsheets.

EQUATIONS

Equations are written in the form

(arithmetic expression) = (arithmetic expression)

where the expressions follow FORTRAN conventions. Some extra facilities provided by SPEEDUP are illustrated in Figure 3. The FOR loop enables sets of equations to be declared easily; the range of the loop extends to the last equation using the FOR loop index. The £ symbol implies a derivative with respect to time. Thus the equation

£HOLDUP(I) = FEED(I) - TOP(I) - BOT(I)

is a differential equation describing the rate of change of the holdup of each component in the drum as a function of feed and offtakes of that component. If a steady-state simulation is being performed, all the derivatives in an equation are set to zero by SPEEDUP. Thus the above equation would become a steady-state material balance on component I.

An alternative way of defining sets of equations to SPEEDUP is using vector arithmetic. The set of equations declared using a FOR loop in Figure 3 could be written as:

BOT = X*BOTT

TOP = Y*TOPT

£HOLDUP = FEED - TOP - BOT

A wide range of matrix-vector arithmetic is permitted.

The last line of the model declares the use of a procedure FLASHV which performs a flash on a mixture whose composition, enthalpy and volume are specified, to determine the amounts and compositions of the two phases and the system pressure and temperature. Function subprograms, such as SQRT, can be used in equations in the same way as in FORTRAN expressions.

2.3 UNIT

The MODEL section describes the models which can be used in the simulation. The UNIT section is used to define which models will be used to simulate particular units, and also declares using SET statements those variables in each unit whose values will be set by the user. Examples of UNIT sections taken from the specification of our sample problem are shown in Figure 4.

```
UNIT D1 IS A FLASHDRUM
    SET VOL,AREA,CV,CL,PTOP
UNIT D2 IS A FLASHDRUM
    SET VOL,AREA,CV,CL,PTOP,PBOT
UNIT FEEDER
    SET X,TEMP,P
```

Figure 4. Examples of UNIT sections.

The first two UNIT sections are those for the two flashdrums in the flowsheet. The third is for a 'unit' FEEDER which is a model allowing the specification of the composition, temperature and pressure of a stream whose total flow is unknown. Note the more compact form of UNIT statement where the MODEL and UNIT names are the same.

2.4 OPERATION

In this section, values are given to all the variables which have been declared as known in the unit SET sections. In dynamic simulations, it is permissible to make these variables arbitrary functions of time, either using standard functions, or using conditional statements. In the latter case, variables can be made to vary depending on what is going on in the simulation. Some examples of OPERATION statements are given in Figure 5.

```
OPERATION
    SET TTIME = 500
    WITHIN D1
        VOL = 0.6,AREA = 0.4,PTOP = 9.9,CV = 0.25,
        CL = 0.015
    WITHIN FEEDER
        X = (0.1,0.15,0.05,0.4,0.3)
        TEMP = 450
        P = IF T>0 and T<200 THEN 55
            ELSE 50 END
```

Figure 5. Examples of OPERATION statements.

The first statement sets a value for the variable TTIME which is the total time to be simulated in seconds. The next group of statements sets constraints for the unit D1. The last group sets values for FEEDER. The pressure is 50 bar at the start of the simulation, increases to 55 bars and remains there for 200 s, and then reverts to its original value. The variable T is the process time in seconds.

2.5 System Implementation

The SPEEDUP system is programmed in PASCAL for ease of production, maintenance and portability and produces a FORTRAN programme. Portability has been demonstrated by the installation of SPEEDUP (which was developed on a CYBER 174) on an IBM 4341 running the IBM PASCAL compiler (VS Release 2) under the UM 370/CMS operating system. The user interacts with a controlling executive, known as SPEEDIT which allows entry and editing of the data and initiation of a run. SPEEDIT manages the files associated with SPEEDUP and controls the sequence of the simulation. The interfaces with machine dependent parameters are dealt with here and are well defined.

Supported by SPEEDIT the translator takes the SPEEDUP description of the problem and produces FORTRAN code which is compiled and then linked with libraries of physical property and numerical routines to perform the simulation. Facilities are provided to halt the execution after a specified time, or at a break point when control returns to the user, who may examine the results so far in graphical or tabular format and then restart the simulation from any chosen time, possibly having changed the values of some variables.

3. Numerical Methods for Steady-State Simulations

In mathematical terms, the steady-state simulation problem can be represented as the solution of a large, sparse set of nonlinear equations. A recent review of the numerical methods available for such problems is given by Sargent (1980).

There are three basic ways of exploiting the sparsity:

(1) by solving the full problem using a numerical method designed for sparse non-linear systems - see, for example Mah and Lin, (1980);

(2) by partitioning the full set of equations and variables into blocks which may be solved sequentially. This can reduce the size of the largest set of equations to be solved simulatenously;

or

(3) by partitioning followed by tearing to reduce the size of the largest subproblem further. Here, a number of variables are guessed, or "torn", in such a way as to allow the remaining variables to be calculated by solving a sequence of single-variable problems. This leaves a set of equations, equal in number to the torn variables, which provide residuals on the basis of which the values of the torn variables may be adjusted in an outer iteration loop.

Algorithms available to perform the analysis of structure to achieve (2) and (3) are described by Sargent (1978).

For the successful implementation of the above options, it is necessary to have available efficient and robust numerical algorithms for the solution of sets of nonlinear equations. In the next section, we describe some work on algorithms which do not exploit sparsity directly. Such algorithms will be of use in implementing options (2) (provided that the subproblems are small) and (3). In section 3.2, some tests on sparse matrix equation solvers (for options (1) and (2) (large subproblems)) are described.

3.1 Methods for full systems of nonlinear equations

Recent numerical tests of the available algorithms for this problem are discussed by Sargent (1980). One of the best recent studies is that of Hiebert (1980). Conclusions from that study are that there is now available a number of robust and efficient codes for solving "well-scaled" problems. HYBRD, an implementation by More and co-workers of Powell's hybrid method, was found to be the most reliable code tested. However, there are problems with these codes in two situations:

(1) where the variables and functions are badly scaled;

or

(2) where the function evaluation is noisy.

We have not addressed the problem of noisy function evaluations. In an effort to overcome difficulty (1), we have developed quasi-Newton methods which have the property of scale-invariance i.e. their behaviour is unaffected by changes in the scale of either the functions or the variables.

The methods are based on the quasi-Newton update of Barnes (1965) and Broyden (1965). The idea is to approximate the Jacobian of a set of nonlinear equations by a matrix B* satisfying

$$y = B^*p \quad (3.1)$$

where y is the difference in function values between the current and last iteration and p is the difference in variable values between the current and last iteration.

A whole class of matrices can be constructed satisfying equation (3.1), using the updating equation

$$B^* = B + (y - Bp)\frac{v^T}{v^T p} \quad (3.2)$$

for arbitrary vectors v (satisfying $v^T p \neq 0$), where B is the approximate Jacobian at the last iteration. Barnes (1965) and Broyden (1965) devised methods based on particular choices of v.

Now, (3.2) is invariant to the scale of the functions, and there are several choices of v which result in quasi-Newton methods which are also invariant to the scale of variables in the problem (Paloschi 1981). Two such choices are:

$$\text{SI1:} \quad v_i = 1/x_i \quad x_i \neq 0 \quad (3.3)$$
$$\phantom{\text{SI1:} \quad v_i} = 0 \quad x_i = 0$$

$$\text{SI2:} \quad v_i = p_i/x_i^2 \quad x_i \neq 0 \quad (3.4)$$
$$\phantom{\text{SI2:} \quad v_i} = 0 \quad x_i = 0$$

where x_i are the elements in the vector of unknowns in the problem.

A numerical comparison of these methods is shown in Table 1. The four methods tested are: HYBRD, the two scale-invariant methods described above, and our own implementation of Broyden's method. The full Hiebert set of 14 problems has been attempted from three starting points: the standard starting point for the problem (x_0); $10x_0$ and $100x_0$. This problem set is labelled A in Table 1. The 42 cases have been repeated with the variables badly scaled using Hiebert's procedure (labelled B), and with the functions badly scaled (labelled C) Measures of robustness and efficiency of the methods are presented in the Table. For robustness, the total number of failures (F) of each method to obtain a solution is reported. For efficiency, a performance ratio (PR), defined as the ratio of the number of iterations taken for each problem to the best achieved by any method for that problem, is computed. Means (MPR) and variances (VPR) of these ratios are presented. Clearly, if the same method was best on all problems, it would achieve an MPR of 1.0 and a VPR of 0.0.

Before discussing the results, we shall describe some details of the implementation of SI1, SI2 and Broyden's method which we have found to be important. First, we update an LU factorisation of the Jacobian approximation, using an algorithm due to Bennett (1965), rather than updating the inverse of B as originally proposed by Broyden. Our numerical results show that this gives codes which fail significantly less often (e.g. 33 as compared to 38 failures of Broyden's method on the Hiebert problem set). Second, we do not require a reduction in the sums of squares of the functions at each step of the methods. We do look at the change in sums of squares over ten iterations as a measure of progress. If the methods fail to reduce the sums of squares by 50% over ten iterations, the Jacobian is reinitialised using finite differences. Last, we use More and Cosnard's (1980) stepping rule to limit the size of each step of the algorithms. Again, we have found that including this rule significantly improves the robustness of all methods. All computations were performed on a CDC 6500 computer in single precision. For more details on the implementation of the methods see Paloschi (1981).

From Table 1, it can be seen that the performance of the scale-invariant methods is comparable to that of HYBRD on the full set of problems. The efficiency of these methods is not quite as good as that of HYBRD, but the scale-invariant methods fail less often. This is because of the performance of the methods does not deteriorate as rapidly when the problem is badly scaled. There are two reasons why the performance of the theoretically scale-invariant algorithms changes when the problem scale is changed. We have not altered the convergence criterion to account for the change of scale, and there are numerical problems with the badly scaled problems which the theory does not take account of, since it assumes infinite precision in all calculations.

One slightly surprising feature of the results is the good performance of Broyden's method on the problems. Although in terms of efficiency its performance is worst of the four methods, it suffers the least number of failures. This observation is in contradiction to the results of Hiebert, who concluded that the implementation of Broyden's method she tested was not competitive with HYBRD. This indicates the importance of implementation details to the performance of the algorithms.

As well as the problems described above, Hiebert also used some chemical equilibrium problems to test the algorithms. Our results for these problems are shown in Table 2. The total iterations to convergence are shown for each method and problem. An entry of zero indicates that no convergence was achieved after 500 iterations. Negative numbers indicate failures detected by the codes. Although the scale-invariant methods (and Broyden's method) perform significantly better on those problems

TABLE 2   Performance of the methods on the chemical equilibrium problems

Problem Number	HYBRD	SI1	SI2	BROYDEN
1	0	-1	-1	-1
2	0	5	5	5
3	0	0	0	0
4	-6	82	104	65
5	-6	14	14	22
6	-6	88	73	68
7	33	0	0	0
8	25	31	26	25
9	-9	62	82	34
10	39	0	0	0
11	-9	0	0	0
12	-9	0	0	0

Failure codes: For HYBRD: -6 : no further improvement possible
                              -9 : argument of function too large
                 For Broyden, SI1, SI2:
                              -1 : singular initial Jacobian

TABLE 1   Results of tests of methods not exploiting sparsity

Method	HYBRD				SI1				SI2				BROYDEN			
Case	A	B	C	Overall	A	B	C	Overall	A	B	C	Overall	A	B	C	Overall
F	7	15	18	40	9	12	14	35	11	13	12	36	9	12	12	33
MPR	1.30	1.25	1.17	1.25	1.59	1.45	1.51	1.52	1.26	1.32	1.55	1.37	1.93	1.83	1.31	1.70
VPR	0.84	1.70	1.14	1.23	2.36	1.30	1.54	1.78	0.72	1.29	2.66	1.56	6.35	4.17	0.93	3.98

Convergence criterion: $\|f\|_\infty < 10^{-7}$

Maximum iterations allowed: 500

than HYBRD, there is still room for improvement!

Overall, the conclusions from these results are:

(1) it is possible to implement algorithms based on scale-invariant quasi-Newton updates which are competitive with the best available nonlinear equation solving codes;

and

(2) there is still scope for improvement of even the best available methods. It is conclusion (2) which is a factor in our decision to allow the inclusion of procedures in SPEEDUP. Specially designed algorithms can then be used for parts of a calculation which are known to cause problems. Work is continuing in an effort to improve the robustness of general purpose methods.

3.2 Methods for solving sparse systems of equations

In this section, we compare the numerical performance of Newton-like algorithms for the direct solution of sparse systems of nonlinear equations. The first method tested is the discrete Newton method, where the Jacobian of the system of equations is estimated by forward difference approximations. The procedure of Curtis, Powell and Reid (1974) is used to reduce the number of function evaluations necessary to evaluate the Jacobian. Two options for the choice of finite difference intervals (h_i) are compared:

Method 1 : $h_i = 0.01\ x_i$

Method 2 : $h_i = 0.01\ f_i$

Method 2 is often called Steffenson's method (Ortega and Rheinboldt, 1970). Method 2 retains the second-order convergence property of Newton's method with analytical derivatives whereas method 1 does not.

Schubert (1970) and Broyden (1971) have presented modifications to Broyden's (1965) quasi-Newton update to take account of the sparsity of the Jacobian. Our implementation of the method based on this update is method 3.

Schubert proposed using his update to change only non-constant non-zero elements in the Jacobian, i.e. elements resulting from nonlinear terms in the equations. Although it is possible to identify linear terms in the equations in a SPEEDUP problem description automatically, obviously it is not worth doing if the performance of the method is not thereby improved. Method 4 is an implementation of Schubert's method based solely on the sparsity pattern of the Jacobian without discrimination between constant and non-constant elements.

The above implementations are all of basic algorithms, without special stepping rules or Jacobian reinitialisation if progress is poor. Method 5 is a modification of method 4 where the Jacobian is reinitialised by finite differences if the norm of the function values is not reduced in ten iterations.

The five methods have been tested on the six problems listed in Table 3. Each problem can be varied in size, and has been attempted for sizes of 50, 100 and 200 variables. Three starting points have been used: the standard one (x_o); $10x_o$ and $100x_o$. Results are presented in Table 4. More detailed results can be found in Bogle (1981).

From Table 4 it can be seen that whilst the quasi-Newton methods are competitive with the Newton methods in terms of efficiency, they fail to converge much more often. There is little to choose between the two Newton methods in terms of both robustness and efficiency, except for starting points close to the solution where presumably the superior local convergence rate of Steffenson's method comes into play.

The most efficient and robust quasi-Newton method is Schubert's original method. Neglecting information on constant Jacobian elements has a deleterious effect on performance. The robustness of the method can be improved by Jacobian reinitialisation if progress is poor, but this causes the efficiency of the method to deteriorate markedly. Overall, it looks as though the quasi-Newton methods are not competitive with Newton's method for sparse problems. This conclusion is in agreement with that of Mah and Lin (1980).

Work is continuing in an effort to improve the robustness of quasi-Newton methods for sparse problems.

TABLE 3   The sparse test problems

Problem Number	Description	N
1	Discrete boundary value problem. Tridiagonal Jacobian, diagonal elements variable.	3
2	Broyden's tridiagonal function Diagonal Jacobian, elements variable.	3
3	Broyden's banded function Jacobian bandwith 7, all elements variable	7
4	Counter-current reactors problem Jacobian bandwith 5, diagonal, sub & super diagonal variable	5
5	Extended Rosenbrock function Block diagonal Jacobian (2x2 blocks). One element in each block variable.	2
6	Extended Powell singular function Block diagonal Jacobian (4x4 blocks). Four elements in each block variable.	4

N : The number of function evaluations needed to compute a Jacobian approximation (neglecting known constant elements)

TABLE 4   Results of tests on sparse equation solvers

	Method	1	2	3	4	5
x_0	F	0	0	6	6	6
	MPR	1.41	1.25	1.00	1.05	1.05
	VPR	0.12	0.03	0.00	0.01	0.01
$10x_0$	F	0	0	9	9	9
	MPR	1.26	1.24	1.06	1.59	1.59
	VPR	0.07	0.06	0.01	0.49	0.49
$100x_0$	F	0	0	9	12	9
	MPR	1.32	1.43	1.18	1.17	2.13
	VPR	0.14	0.19	0.07	0.03	2.10
Overall	F	0	0	24	27	24
	MPR	1.33	1.31	1.07	1.26	1.33
	VPR	0.11	0.09	0.02	0.18	0.78

Convergence criterion : $\|f\|_\infty < 10^{-7}$

Maximum iterations allowed : 200

3.3 Decomposition methods for systems of equations

We have also studied methods for decomposing large problems by partitioning and tearing. In earlier work it seems to have been assumed that partitioning is always worthwhile, and if tearing is used, that the optimum strategy is to use a minimum number of torn variables. Hernandez and Sargent (1979) attempted to assess the relative merits of partitioning and tearing, and suggested an extension of the tearing technique to obtain a sequence of either single-variable problems or linear subsystems.

Our tests show that the minimum-tear strategy often leads to extremely bad conditioning in the outer-loop, and that the linear subsystems of the extended strategy can also be ill-conditioned and frequently even functionally singular. Work has therefore been directed to investigating alternative criteria for selecting torn variables.

Clearly a direct assessment of conditioning of the subproblems is impractical, and the most promising approach seems to be to attempt to minimize interaction between the subproblems. Thus the torn variables are chosen so that they have a minimal effect on the determination of the remaining variables, which in turn have a minimal effect on the outer-loop residuals. Similarly, if linear subsystems are allowed they are restricted to being diagonally dominant. Further details, together with numerical results, are given by Edwards (1981).

The limited amount of testing carried out so far shows that decompositions based on this technique are considerably more robust than those given by a minimum-tear strategy, but not as robust as direct methods for solving the undecomposed system without taking account of sparsity. Similarly, because of the time required for the analysis of the system, they do not compete in efficiency either.

However, tests on large sparse problems to compare with the direct methods for such systems described in Section 3.2 above have yet to be carried out, and the outcome is still an open question.

4. Numerical Methods for Dynamic Simulation

The general transient problem results in

a set of coupled ordinary differential equations and nonlinear algebraic equations. The two basic approaches to solving these coupled systems is dictated by the two classes of integration methods, i.e. explicit and implicit.

Explicit methods do not require any iterative procedure to advance the solution of the differential variables. However, the coupled algebraic system must be solved for each evaluation of the derivatives using some type of quasi-Newton method. Such methods as explicit Runge-Kutta and linear multistep (Adams) methods fall into this class. For neutrally stable, non-stiff problems these methods are computationally reliable and efficient, provided automatic error control is used to adjust the step-length.

For stiff problems, implicit methods are required because of their increased stability. The most obvious class of methods are the backward differentiation formulae (BDF) proposed by Gear (1971). These can be formulated to solve the differential-algebraic equation problem and have proved very effective.

Another class of methods being considered is due to Nørsett (1974). These are the semi-implicit (explicit) Runge-Kutta (SIRK) methods. The methods can possess very strong stability properties which makes them superior to the BDF under certain circumstances. Cameron (1981) has shown that they can be implemented in fixed order-variable step or variable order-variable step modes, and has developed an A-stable variable-order code, but the benefit of variable-order implementation is often not fully realized because of the extra overhead required in this mode.

Numerical experience suggests that the SIRK methods are competitive with BDF methods at moderate to low local error tolerances (10^{-2} to 10^{-4}) as normally used in engineering applications; for high accuracy however the BDF prove superior. Also, the SIRK methods generally require more evaluations of the equations than the BDF, and hence when the equations include extensive subroutines the BDF methods become increasingly superior. On the other hand, for startup of discontinuous problems the SIRK methods can maintain high order where the BDF methods must reduce order and reinitialize at the point of interruption. In these circumstances special high order starting procedures are required for BDF.

Because of the extent of the stability regions of implicit methods, stable solutions may be generated for truly unstable problems. This must be appreciated in using these methods, and in the case of unstable problems explicit methods are preferred since they usually detect solution growth. Techniques are being sought to overcome this defect in implicit methods.

5. General Conclusions

Good general-purpose methods are available for solving the underlying numerical problems for both steady-state and dynamic simulation. With reasonable initial estimates based on physical intuition, these should produce solutions in the majority of cases and the ordinary user should not need to be concerned with them.

However, a general-purpose simulation system can be expected to throw up problems from time to time which will be a severe test for any method, and the system must therefore provide alternative methods, with sufficient diagnostic information to guide the choice for more knowledgeable users. There is much scope for improving such diagnostics, and improving robustness of the software by building in automatic corrective action.

For solving sets of nonlinear algebraic equations, the most robust methods are not necessarily the most efficient, and further study of modes of failure could produce worthwhile improvements.

Improvements in solving systems of algebraic equations will also improve the methods for solving coupled differential and algebraic systems. Unfortunately, for such systems there is not a universal method which is best for all problems, although in most cases use of an inappropriate method merely affects the efficiency. A major difficulty is to detect unstable behaviour in a system which in normal circumstances exhibits stiff characteristics, and further work is required for this class of problem.

LITERATURE CITED

1. Barnes, J.G.P., "An Algorithm for Solving Nonlinear Equations Based on the Secant Method", Computer J., 8, 66-72 (1965).

2. Bennett, J.M., "Triangular Factors of Modified Matrices", Numer. Math., 7, 217-221 (1965).

3. Bogle, I.D.L., "A Comparison of Algorithms for Solving Large Sparse Systems of Nonlinear Equations", Imperial College Report (1981).

4. Broyden, C.G., "A Class of Methods for Solving Nonlinear Simultaneous Equations", Math. Comp., 19, 577-593 (1965).

5. Broyden, C.G., "The Convergence of an Algorithm for Solving Sparse Nonlinear Systems", Math. Comp., 25, 285-294 (1971).

6. Cameron, I.T., "Numerical Solution of Differential-Algebraic Systems in Process Dynamics", PhD Thesis, University of London (1981).

7. Curtis, A.R., M.J.D. Powell and J.K. Reid, "On the Estimation of Sparse Jacobian Matrices", J.Inst.Math.Appl., 13, 117-120 (1974).

8. Edwards, D.W., "Robust Decomposition Techniques for Process Design and Simulation", PhD Thesis, University of London (1981).

9. Gear, C.W., "Simultaneous Numerical Solution of Differential-Algebraic Equations", IEEE Trans.Circ.Theory CT-18(1), 89-95 (1971).

10. Hernandez, R., and R.W.H. Sargent, "A New Algorithm for Process Flowsheeting", Computers and Chemical Engineering, 3, 363-371 (1979).

11. Hiebert, K.L., "A comparison of Software Which Solves Systems of Nonlinear Equations", Sandia Technical Report SAND80-0181, Albuquerque (March, 1980).

12. Mah, R.S.H., and T.D. Lin, "Comparison of Modified Newton's Methods", Computers and Chemical Engineering, 4, 75-78 (1980).

13. Malathronas, J.P., and J.D. Perkins, (1980) "Solution of Design Problems Using Broyden's Method in a Sequential-Modular Flowsheeting Package", presented at "CHEMPLANT 80" Heviz, Hungary, September 1980.

14. More, J.J., and M.Y. Cosnard, "Numerical Solution of Nonlinear Equations", ACM Trans. Math.Software, 5(1) 64-85 (1979).

15. Norsett, S.P., "Semi-explicit Runge-Kutta Methods", Rpt.6/74, Dept. of Mathematics, University of Trondheim, Norway (1974).

16. Ortega, J.M., and W.C. Rheinboldt, "Iterative Solution of Nonlinear Equations in Several Variables", Academic Press, New York (1970).

17. Paloschi, J.R., "Scale-Invariant Quasi-Newton Methods for sets of Nonlinear Equations", Imperial College Report (1981).

18. Sargent, R.W.H., (1981), "The Decomposition of Systems of Procedures and Algebraic Equations", in G.A. Watson (Ed.) "Numerical Analysis - Proceedings, Biennial Conference, Dundee, 1977", Lecture Notes in Mathematics, 630, 158-178, Springer-Verlag (Berlin, 1978).

19. Sargent, R.W.H., (1981), "A Review of Methods for Solving Nonlinear Algebraic Equations", in R.S.H. Mah and W.D. Seider (eds.), "Foundations of Computer-Aided Chemical Process Design", Vol. 1, pp 27-76, Engineering Foundation (New York).

20. Schubert, L.K., "Modification of a Quasi-Newton Method for Nonlinear Equations with a Sparse Jacobian", Math.Comp., 24, 27-30 (1970).

DEVELOPMENT OF A NEW EQUATION-BASED PROCESS FLOWSHEETING SYSTEM: NUMERICAL STUDIES

MARK A. STADTHERR
Chemical Engineering Department
University of Illinois
Urbana, IL 61801

and

COURTLAND M. HILTON
Intel Corporation
Aloha, Oregon 97006

A new equation-based flowsheeting system is described. The system uses a simultaneous linearization approach and employs powerful sparse matrix routines. The equation generation procedure is efficient and makes flowsheet input easy. The flowsheet input format is very much like the format used in conventional sequential modular simulators. The current system is a prototype that is being used to study a number of fundamental computational problems. Results of such numerical studies are reported.

Current methods for chemical process simulation and design (flowsheeting) are typically based on the sequential-modular approach, in which the computations for each type of unit operation are organized into modules and solved sequentially. However, this approach has inherent limitations that make it ineffective in dealing with large and complex processes. Since processes of the future are likely to be increasingly complex, as various schemes for conserving material and energy and for controlling effluents are implemented, the development of effective flowsheeting systems for such processes is particularly timely. One strategy for overcoming the current limitations is the equation-based approach. In this case, the computational modules are done away with and all the equations describing a process solved simultaneously. One approach to solving the very large equation system that results is algebraic decomposition (or tearing). In this case, the system may be solved by iterating on relatively few variables. Though this intuitively seems desirable, serious computational problems can arise, as noted by Lin and Mah (1978), who advocate a simultaneous linearization approach as potentially more powerful. In this case, all the equations are linearized and all variables iterated on simultaneously using a Newton-Raphson or quasi-Newton approach. Although in the past, equation-based flowsheeting calculations were based primarily on algebraic decomposition, there has been considerable recent interest (Gorczynski and Hutchison, 1978; Gorczynski et al., 1979; Westerberg and Berna, 1978; Lin and Mah, 1978; Benjamin et al., 1981) in the simultaneous linearization approach. This is the approach used here. The new system is a prototype that is being used in studies of a number of fundamental computational problems. In this paper, we describe the prototype system and present results for some of the studies made using the system.

One study reported here involves the reliability with which convergence is obtained from the "worst-case" initial guess situation. This is the case in which the user supplies no initial guess values and all must be generated internally by the program. We compare the convergence behavior for three schemes for internally generating an initial guess, and for several strategies for linearizing the nonlinear equations. Also presented is a study involving the reliability of convergence from a "best-case" initial guess situation using several different linearization strategies. Different strategies for computing thermophysical properties are also considered.

BACKGROUND

Extensive reviews of past work in the area of process flowsheeting have been provided by Motard et al. (1975) and Hlavacek

---

This work was performed at the University of Illinois. Correspondence should be addressed to Prof. Stadtherr.

(1977), and more recent work has been reviewed by Rosen (1980), and Evans (1981). These reviews discuss many of the advantages and disadvantages of the modular and equation-based approaches to process flowsheeting. Also, the recent monograph of Westerberg et al. (1979), is particularly useful as an introduction to the field of process flowsheeting.

One can distinguish between two general approaches to equation-based flowsheeting. The first of these, typified by the SPEEDUP system (Leigh et al., 1974), involves tearing. One guesses, or tears, values for a number of variables sufficient to permit values for the remaining variables to be found by solving a sequence of small, usually one-variable problems. The remaining equations (tear equations) may then be solved for new values of the tear variables, and some sort of successive substitution procedure used, providing of course that the tear equations contain the tear variables explicitly. If this is not the case, the residuals in the tear equations can be used in connection with other standard root-finding procedures. Thus in effect one is able to solve a large system of nonlinear equations by iterating on only a few tear variables, thereby drastically reducing the dimensionality of the problem.

The key step in the tearing approach is the development of an appropriate solution strategy (information flow pattern) for the particular problem at hand. That is, one must decide which variables to tear, which equations to solve for which variables (output set), and in which sequence to solve them (precedence order). Furthermore, since most equation systems describing chemical processes will be underconstrained, it is also often necessary to designate certain variables as design variables. What is needed then is a systematic and efficient procedure for finding a solution strategy that will converge reliably and rapidly to the solution. Over the past several years, a variety of techniques have been devised with this as a goal. Many are designed to produce a solution strategy involving a minimum number of tear variables. Others also try to account for the relative difficulty of the single-variable problems that must be solved. Still others involve sensitivity considerations, for reasons discussed in more detail shortly. Many such techniques are prone to combinatorial problems and thus are not reliably efficient. Moreover, there is little degree of certainty that the solution strategies obtained will reliably converge. Thus, a basic premise of the tearing approach, namely that a small-dimensioned problem is easier to solve than a large-dimensioned one, is not necessarily correct. This has been emphasized by Lin and Mah (1978), who point out that because a very long chain of computations typically exists between the guessed tear values and the residuals in the tear equations, sensitivity problems can arise that may cause divergence even for initial guesses very near the solution. For this reason, methods for employing sensitivity considerations in choosing a solution strategy have been developed, as mentioned above. Compared to the sequential-modular approach, the tearing approach is typically faster and capable of solving more complex flowsheeting problems. Nevertheless, because the problem of efficiently choosing a reliable solution strategy has not been completely solved, there has been reluctance to adopt this approach.

A second and more promising approach to equation-based flowsheeting is the quasilinear approach. This involves the simultaneous linearization of all the equations and iteration on all the variables, using the Newton-Raphson method, a quasi-Newton method, or some hybrid thereof. Thus in each iteration we are faced with solving a huge set of sparse linear equations, involving perhaps several thousand variables. At this point the quasilinear approach has been proven very successful in dealing with some specialized problems, such as flows in pipe networks (Mah, 1974; Bending and Hutchison, 1973) and simulation of distillation columns (Hutchison and Shewchuk, 1974; Kubicek et al., 1976). These applications have been summarized by Westerberg et al. (1979), who also emphasize the promising aspects of this approach. Application of the quasilinear approach to flowsheeting problems in general is a very recent development. The work of Mah and Lin (1978), who apply this approach to a flowsheeting problem involving the simulation of a natural gas liquefaction process, is indicative of the potential of this approach, but they do not provide a generalized flowsheeting system. Hutchison and coworkers at the University of Cambridge in Great Britain have recently described a flowsheeting package called QUASILIN (Gorczynski et al., 1979) that may be regarded as a prototype of such a generalized quasilinear flowsheeting system. The ASCEND II package (Benjamin et al., 1981) developed by Westerberg and co-workers at Carnegie-Mellon is equally noteworthy in this regard.

The most fundamental computational problems with the quasilinear approach involve the strategy to be used to converge the nonlinear equations, and the strategy to be used to solve the huge sparse linear systems that arise after linearization. The sparse matrix strategy used determines in effect the limit on the size of problems that can be solved. For instance, consider the problem of solving the linear system $Ax = b$. The usual solution procedure can be represented by the factorization $A = LU$ where L is lower triangular and U is upper triangular. The elements of L and U are usually found using Gaussian elimination or some variation thereof. If A is large and sparse, the number of nontrivial nonzeros in L and U may greatly exceed the number of nonzeros in A. This loss of sparsity, or "fill-in", may lead to excessive storage requirements and unacceptably long computational times. Without the use of sparse matrix strategies to reduce this fill-in, only very small flowsheeting problems could be handled by the quasilinear approach. Since one of the main reasons for adopting an equation-based approach is its ability to handle large and complex problems, the need for effective sparse matrix strategies is particularly important. Westerberg et al. (1979) emphasize that while current sparse matrix software is capable of handling problems involving one or two thousand equations and sometimes more, realistic chemical plant flowsheeting problems will require sparse matrix strategies capable of handling equation systems larger by an order of magnitude or more. An important feature of the new flowsheeting system we are developing is that it is interfaced to a set of powerful new sparse matrix routines that can handle problems involving several thousand equations without resort to decomposition. This represents significant progress toward the goal cited above.

The strategy used to converge the nonlinear equations determines, to a great extent, the reliability and speed with which a given problem can be solved. Straightforward Newton-Raphson is not particularly attractive because, as discussed below, it may be difficult to supply a good enough initial guess. There are a variety of approaches to improving convergence from a poor initial guess. One promising approach is that described by Gorczynski and Hutchison (1978). They note that since a second-order linearization method, such as Newton-Raphson, provides speed at the expense of reliability, while first-order methods provide reliability at the expense of speed, a blend of the two linearization strategies would seem to be appropriate. Indeed they seem to have had some success with such blended linearizations. Unfortunately, there is little published regarding such important details as the actual values of the blending parameters used. Another approach to the convergence problem is to use "hybrid" or "dogleg" methods (Powell, 1970; Westerberg and Director, 1978) that consider the steepest-descent direction in addition to the Newton-Raphson direction. A hybrid method recently described by Chen and Stadtherr (1981) appears to be very reliable. Though the current code based on this method is written for full matrix problems, it can also be extended to sparse systems of nonlinear equations. This work is currently in progress. The need for nonlinear equation solvers capable of converging from poor initial guesses could of course be ameliorated considerably given an efficient procedure for generating a good initial guess. The difficulty here is that since all variables are iterated on, they all require initial values. For relatively small problems, the user may be quite capable of supplying a good initial guess; however, on larger problems, the user may not be able to provide initial guesses for all of the several thousand variables that may be involved. Thus one needs algorithms for systematically generating "good" initial guesses given little or no user input. Gorczynski and Hutchison (1978) outline a simple initialization scheme, though again little numerical detail is provided, and it is unclear how effective it is.

DESCRIPTION OF FLOWSHEETING SYSTEM

The system described here, which we have dubbed SEQUEL, is a prototype of a new equation-based process flowsheeting system. As the name implies, SEQUEL is an outgrowth of previous work in process flowsheeting and represents a new chapter in the continuing effort to design and simulate chemical processes by equation-based methods. It must be emphasized that SEQUEL has been designed primarily as a tool for use in developing and evaluating computational strategies for equation-based flowsheeting. Thus, SEQUEL is lacking several features that would be important in a production code. For instance, although flowsheet input to SEQUEL is very easy, friendliness to the user has in general been a secondary concern. Also, the number of components in the physical property data base is quite small and the thermodynamic models used are very simple and are inadequate for

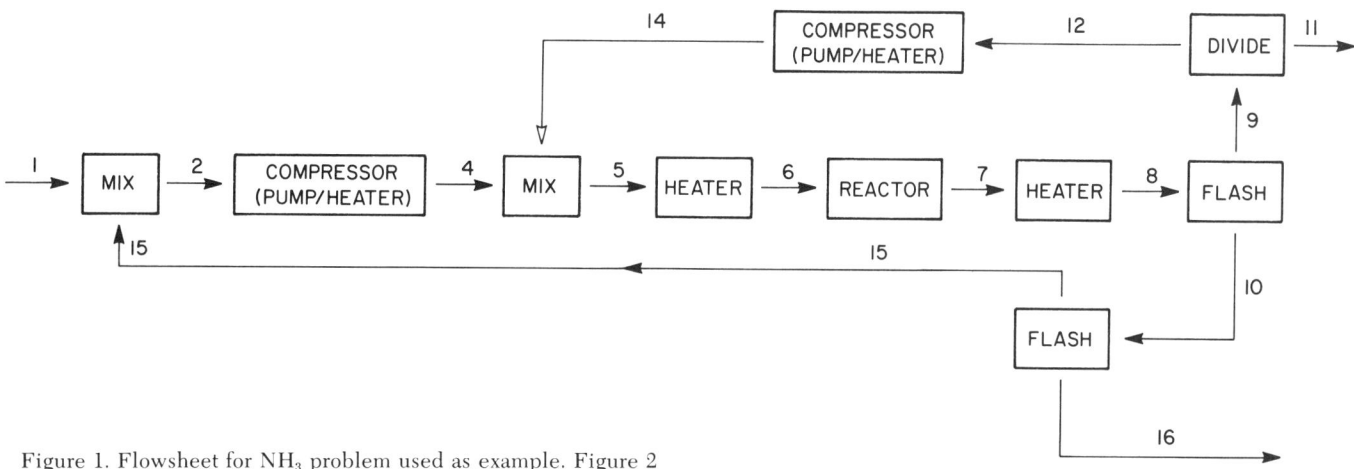

Figure 1. Flowsheet for NH$_3$ problem used as example. Figure 2 shows the flowsheet input file for this flowsheet.

some problems.

Equation Generation

One problem commonly cited (e.g., Evans, 1981) in connection with the equation-based approach is that a complex executive routine is needed to generate the equations describing a particular flowsheet. Two ideas have appeared in recent years that aid in overcoming this difficulty. The first such idea is modularity (Gorczynski et al., 1979; Benjamin et al., 1981). SEQUEL is organized in a modular fashion, with the modules corresponding to the various standard process units. Of course in this context, the modules are not procedures for equation solving, as in the sequential-modular case, but instead are used for generating the equations for a particular process unit. The modular architecture is a natural response to the need for flexibility and ease of user input, as well as the desire for an efficient and easily understood code. The second idea (Gorczynski et al., 1979; Mah and Lin, 1978; Book and Ramirez, 1978) for simplifying the equation generation code involves the observation that the equations representing a process, typically involve a rather limited number of equation types. In SEQUEL, this is exploited by using a "library" of standard equation types. It is assumed that the flowsheet can always be represented by a set of equations of these standard types. When necessary, the library of standard types can be augmented by the user. The use of standard equation types is a very efficient way to store the information required for Jacobian and function evaluation. Rather than maintain a copy of this information for each occurrence of a particular equation type, only one copy is required. It is also worth noting that by adopting this approach, we avoid the use of algebraic equation-manipulation pro-

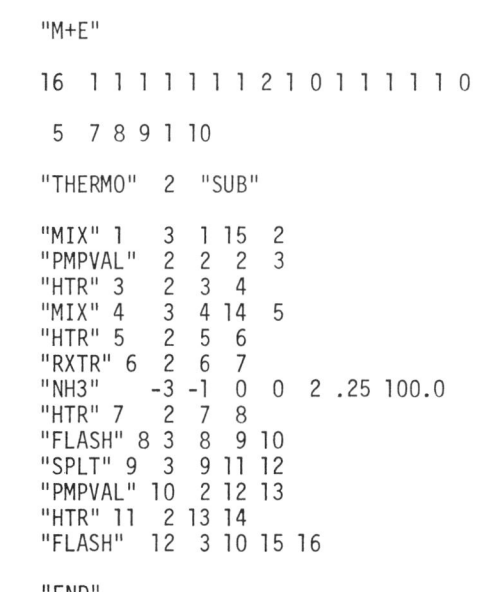

Figure 2. Flowsheet input file for NH$_3$ problem shown in Figure 1. See text for additional detail.

grams, whose use in this context has sometimes been proposed (Hanyak, 1980; Coup et al., 1981; Kubicek et al., 1976).

Flowsheet input to the equation generator uses standard FORTRAN free-format conventions. Figure 1 is an example of a chemical process flowsheet; the input file describing this flowsheet is shown in Figure 2. There are six types of statements in the input file:

1. The first statement specifies whether the equation set will describe only a material balance of the process or a complete material and energy balance.

2. The second input is the total number

of streams in the flowsheet together with a specification of the phase of each stream. The user may specify the stream as all vapor, all liquid, or of unknown state.

3. The third input is the number of chemical species in the process together with their property data bank identification numbers.

4. The fourth gives the type of thermodynamic model to be used for predicting vapor-liquid equilibria, and whether the thermophysical properties are to be evaluated in external subroutines or as part of the overall matrix.

5. Fifth, the unit operations involved are listed followed by their user identification number, the number of streams connected to the unit operation, and the numeric designator of each such stream. The unit operations may be listed in any order.

6. Sixth, the input list is terminated with an END statement.

Like in many other flowsheeting programs, each connecting stream i is described by a vector S_i comprising molar component flowrates, total stream flowrate, temperature, pressure, and total stream enthalpy. That is, $S_i = (n_i, F_i, T_i, P_i, H_i)$. Component flowrates were chosen as primitive variables rather than mole fractions because material balances around a unit could then be represented by linear equations. The total stream flow was included for clarity and to simplify the structure of some equations. Of course, the inclusion of both component flows and total stream flow as variables requires an additional equation for each stream, namely the stream balance that equates total flow to the sum of the component flows. After reading the number of streams and number of components, SEQUEL creates a variable vector X that comprises all of the stream vectors. That is $\underline{X} = (\underline{S}_1, \underline{S}_2, \ldots, \underline{S}_M)$ where M is the number of streams. Since the order of the variables within each stream vector is the same, it is easy to locate any individual variable within the X-vector. After initialization, the X-vector contains only the external variables describing flows between units. As discussed below, internal variables are appended to the X-vector as they arise in generating the equations.

As each unit operation in the input file is read, control within the equation generator branches to the module for that type of unit

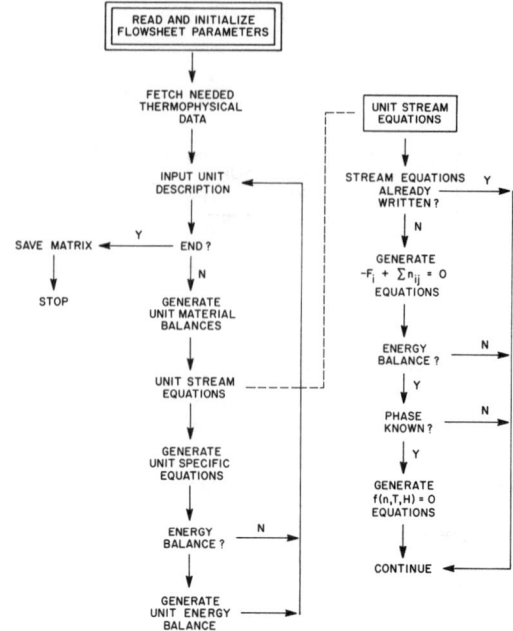

Figure 3. Information flow in equation generation routine.

operation. This module subsequently issues calls to a subroutine library that creates general balance equations as well as an equation set unique to the unit operation and stream configuration specified. Figure 3 illustrates the general program flow during equation generation.

The equations representing any given unit operation are a subset of a "library" of standard equation types, as shown in Figure 4. In defining standard equation types, care must be taken to structure the equations so that numerical difficulties such as division by zero are avoided. Additional equation types can be added as necessary.

As the equations are generated, occurrence matrix information is stored by sequentially filling an array CI with the position in the X-vector of each variable contained in an equation. A pointer array RLI is maintained that indicates where in CI each row (equation) begins. An additional vector EQN is prepared that contains numerical values specifying the equation type corresponding to each row of the occurrence matrix. If an equation type involves coefficients that vary from row to row, these constants are stored in another vector. As an example of the equation generation process, we show in Figure 5 the arrays produced when the material balances around a heater are generated.

Some equations involve the use of internal variables such as K-values, split fractions, and heat loads. As these variables

Equation Type		Equation Type	
1	$-X_0 + \sum_i X_i = 0$	10	$-X_0 + \sum_{i=1}^{N} b_i X_i + \sum_{i=1}^{N} X_i c_{pLi}(c_i - c_0) + \sum_{i=1}^{N} X_i \int_{c_i}^{X_{N+1}} C_{pVi} dX_{N+1} = 0$ (vapor)
2	$-X_0 + c = 0$		$-X_0 + \sum_{i=1}^{N} X_i C_{pLi}(X_{N+1} - c_0) = 0$ (liquid)
3	$-cX_0 + X_1 = 0$		where:
4	$-X_0 + \sum_i X_i + c = 0$		$C_{pLi} = a_i \left[d_1 + d_2(1 - \frac{X_{N+1}}{e_i})^4 + d_3(1 - \frac{X_{N+1}}{e_i})^{-1} \right]$
5	$\sum_i X_i + c = 0$	11	$-X_0 X_1 + \exp\left[c_1 + \frac{c_2}{X_2} + c_3 \ln X_2 + c_4 \frac{X_0 X_1}{X_2^2} \right] = 0$
6	$-X_0 + \min[X_1, X_2, \ldots, X_i] = 0$		
7	$-X_0 X_1 + X_2 = 0$	12	$-X_0 + X_1 + c_0 \min\left[\frac{X_1}{c_1}, \frac{X_2}{c_2}, \ldots \right] = 0$
8	$-X_0 X_1 X_2 + X_3 X_4 = 0$		
9	$-X_0 X_1 + \exp\left[c_1 - \frac{c_2}{c_3 + X_2} \right] = 0$	13	$-X_0 + X_1 + X_2 + c_0 \min\left[\frac{X_3}{c_3}, \frac{X_4}{c_4}, \ldots \right] = 0$

Figure 4. Partial list of standard equation types used in SEQUEL. There are currently 22 standard equation types. The X_i's are the variables; the c's are constants.

arise in generating the equations they are sequentially assigned a position in the X-vector. An identification vector is also maintained containing the type of variable at each position for use when subjecting the X-vector to global constraints, as discussed below.

The arrays produced by the equation generator are used directly when performing function evaluations and when creating and updating the linearized model at each iteration. When a call is issued by the solution portion of SEQUEL to evaluate functions or derivatives, EQN is sequentially scanned. For each equation I, the program branches to the procedure specified by EQN (I), and using the occurrence matrix information in CI in conjunction with the values in the X-vector, the desired expressions are evaluated. As previously stated, this type of program, in which each procedure (equation type) is coded only once, results in a tremendous reduction in core requirements, and makes equation generation efficient even for very large problems.

Thermophysical Property Models

As noted above, the models available to date in SEQUEL are very simple and will not always be adequate. In part, this is because our interest here is less in the accuracy of the models than in the computational strategy for using them. For determining such properties as K-values, dew and bubble points, enthalpies, vapor fractions, etc., one can in general identify three types of computational strategies:

Case 1. In the simplest case all thermophysical properties are evaluated in external subroutines. Therefore, the thermophysical models are not treated as part of the overall process matrix and are not subject to linearization. That is, no linearized K-value, dew and bubble point, or phase determination equations appear in the process matrix. A linearized enthalpy equation for each stream in the flowsheet is included in the matrix to allow the update of each stream's temperature. After a process matrix iteration, current values of temperature, pressure, and composition are provided to the subroutines, which in turn update, perhaps iteratively, the thermophysical property values required for function and Jacobian evaluation. The omission of the thermophysical property expressions from the matrix keeps the process matrix size smaller than would otherwise be possible. However, except for stream enthalpies, no information on the variation of the property values with temperature, pressure or composition is included. A linearized process model based on such a strategy is inaccurate and its convergence behavior may be poor for some problems. For example, in modeling a process

$S_1 \rightarrow \boxed{Q} \rightarrow S_2$

		Variable Slate	Position in X-vector S_1	S_2
n_{11}	n_{21}	n_1	1	7
n_{12}	n_{22}	n_2	2	8
F_1	F_2	F	3	9
T_1	T_2	T	4	10
P_1	P_2	P	5	11
H_1	H_2	H	6	12

Material Balances:

$-F_1 + F_2 = 0$

$-n_{11} + n_{21} = 0$

Stream Balances:

$-F_1 + n_{11} + n_{12} = 0$

$-F_2 + n_{21} + n_{22} = 0$

in terms of X-vector components:	Equation type
$-X_3 + X_9 = 0$	1
$-X_1 + X_7 = 0$	1
$-X_3 + X_1 + X_2 = 0$	1
$-X_9 + X_7 + X_8 = 0$	1

Vector	Position 1	2	3	4	5	6	7	8	9	10
EQN	1	1	1	1						
RLI	1	3	5	8	11					
CI	3	9	1	7	3	1	2	9	7	8

Figure 5. Arrays created when generating material and stream balance equations around a heater unit.

including an adiabatic flash, it is important to include the K-value dependence on temperature in the linearized process model. A model without this information, in general, will exhibit poor convergence behavior. On the other hand, if one were modeling a process involving only isothermal flashes, good convergence could still be expected.

Case 2. This case is like the first, with the distinction that linearized K-value equations are now included in the process matrix. This contributes to a possibly drastic increase in the size of the occurrence matrix, since multiple copies of the linearized K-value model must be included. However, the model is more accurate and may be expected to perform well in both the adiabatic and isothermal cases. The subroutines are used to update the property values, and may also be used to generate the coefficients in the linearized models. This means that within each Newton-Raphson iteration, there may be nested iteration loops in the subroutines that need to be converged.

Case 3. A third type of strategy avoids this nesting of iteration loops by eliminating the subroutines and treating the physical property models simply as equations within the overall set whose residuals are to be driven to zero by the Newton-Raphson process. In this case, the thermophysical property models are not converged until the end, and the properties are updated at each iteration by the Newton-Raphson process and not by external subroutines. In comparison to Case 2 this may mean another drastic increase in the size of the occurrence matrix. We are currently using SEQUEL to perform numerical studies comparing these three strategies.

Though the models used to date are simple, the flexible nature of SEQUEL allows the inclusion of more complex models with little difficulty. Thermodynamic models are represented as standard equation types within the library set. It should be noted that a standard equation type may in fact be a procedure involving several equations. Thus, complex models can be handled in this context. In the current version of SEQUEL, all streams, both liquid and vapor, are modeled as ideal solutions. Raoult's law is used for K-value prediction, along with vapor pressures predicted using either the Antoine equation or an equation of standard type 11 as given by

Reid, Prausnitz, and Sherwood (1977). The principal differences between the vapor pressure equations is that they are, respectively, explicit and implicit in vapor pressure, the latter thus requiring iterative solution. Heat capacities of vapor components are evaluated using third order polynomials in temperature. Liquid component heat capacities are evaluated using an equation of standard type 10. The thermophysical property data base is a subset of the data base included in Reid, Prausnitz, and Sherwood (1977) and thus is very similar to the CACHE data base. The data base is stored in a separate file and is accessed only once.

Design Specifications

After the equation generation routine has processed the flowsheet input file, the equation set generated will in general contain more variables than equations. Thus, design variables must be chosen and assigned values, or other specifications added to the equation set. Design variables are handled by adding to the equation set equations that set the design variables equal to their specified values. In this version of SEQUEL the user is required to choose the design variables and supply a consistent set of specifications. Though the user entry of this information is straightforward, the program could be made friendlier to the user by incorporating one of the procedures available for automatically selecting some or all of the design variables.

Initialization

Since all variables are iterated on, they all require initialization. Ideally these initial guesses would be provided by a user whose insight into the process at hand permits him to provide good guesses. For relatively small problems, this is not an unreasonable expectation. For larger problems, however, this may be impractical. Nevertheless, even on large problems a relatively good initialization may be available. For instance, the result of a similar problem solved earlier may be used. Also, by starting with a small part of the overall process for which a good guess can be made, one may exploit the modular nature of SEQUEL to "build" a process a unit or two at a time. This is essentially a "continuation" approach in which a sequence of problems is solved, the solution of one problem being used to generate initial guesses for the next.

The worst case with respect to initialization is that the user will be unable to supply an initial guess, requiring the program to assign initial values internally. In our numerical studies using SEQUEL one interest is in studying the performance of the quasilinear approach in this situation. Though any number of internal initialization schemes might be devised, for the purpose of the numerical studies presented here we have used three such "worst-case" schemes. In all three schemes, temperature, pressure, and enthalpy are set to preselected reference values, heat loads are set to zero, and split fractions are set equal. These values are overridden if other values are specified as design variables. The guesses for the K-values are calculated using these preset or specified temperatures and pressure values. The three schemes differ in how component flowrates are guessed:

Type 1. In this most primitive scheme all unspecified component flows are set to some arbitrarily small number (we used 0.001). Somewhat surprisingly this works remarkably well for some kinds of problems.

Type 2. In this case all component flows are set to some arbitrary "average" values. As in the previous scheme each stream will have the same component flowrate values, except where overridden by a design specification. There are any number of ways for choosing the "average values". For instance, one could simply use specified or estimated feed stream values.

Type 3. In this case a set of heuristics is used, one for each type of unit, and a guess is generated by beginning with the feed stream and moving sequentially through the process until guessed flows for all streams have been generated. Only one pass is made through the process, so there is no iteration on recycle streams. The heuristics are as follows:

1. FEED STREAMS. Feed component flows not specified are estimated using specifications for those components elsewhere in the flowsheet if possible. For a stream component for which this is possible, the feed flow is set to the average of flows specified elsewhere for that component times the number of output streams. For components that do not have a flow specified anywhere in the flowsheet, the feed stream flows are set to the average of all specified component flows.

2. STREAM MIXER. Guess the output stream by assuming that any unknown inlet flows contribute one-half the flow in the largest known (or already guessed) inlet.

3. STREAM DIVIDER. Guess outlet streams using specified outlet/inlet ratios if available. Otherwise assume inlet flow is divided equally among outlets.

4. EQUILIBRIUM FLASH. Guess outlet streams by assuming total liquid flow out equals total vapor flow out. Use the K-values guessed as described above.

5. REACTOR. If conversion is not specified, guess outlet by assuming 100% conversion of limiting reactant in inlet.

In using these heuristics, values given by design specifications always override the guessed flows and assumed parameters. Many other heuristics such as these could be used, and may be better than those given here.

Equation Linearization and Solution

Three linearizations are currently available in SEQUEL. These are: standard second-order Newton-Raphson linearization, Newton-Raphson with step size relaxation, and a hybrid linearization that blends first and second order linearizations according to user specification. The hybrid linearization, an approach suggested by Gorczinski and Hutchison (1978) is based on _a priori_ analysis of the nonlinear equation types. Each nonlinear term is linearized by selecting all but one of the variables in the term to be constant. The selection was made when SEQUEL was programmed. The resulting linearization is first order. It is then straightforward to gradually shift from the first order representation to the full Newton-Raphson linearization. The user specifies the value of the blending parameter (zero to one, where one is a full Newton-Raphson linearization) and the rate at which the blending parameter is to be increased. The idea, of course, is that far from the solution the stability of the first-order linearization is desirable, while nearer the solution the speed of the second-order linearization is desirable. While requiring the user to select the blending parameter and rate may not be desirable in a production code, the requirement is consistent with our use of SEQUEL to study different computational strategies.

X-vector values may vary considerably from iteration to iteration, and values may be outside the range of physical feasibility, e.g., negative flowrates. In some cases, this may cause no difficulty, but in others it may create numerical difficulties in performing function and derivative evaluations. The values for the X-vector are therefore constrained to be within user-defined bounds of feasibility. SEQUEL utilizes global constraints as opposed to constraints on individual variables. For example, all flowrates are subject to the same constraints, rather than varying flowrate constraints from component to component and stream to stream. SEQUEL constrains pressures and flowrates to be positive, and, to prevent excursions too far from the applicable range of thermophysical property models, constrains temperatures within maximum and minimum values and enthalpies below some maximum value. When a constraint is violated, the X-vector values are reset heuristically. Convergence is monitored using the norm of the correction vector as well as the norm of the residual vector.

Sparse Matrix Processing

For solving the linearized equations, we use a two-pass approach. In the first pass, the equations and variables are reordered into a matrix form that reduces fill-in during the solution. In the second pass, the linear system is solved using Gaussian elimination or some other technique. In this case, since all the linearized problems have the same structure, the matrix reordering is done only once and is saved for use in later iterations.

Various reordering and solution methods are available to SEQUEL. For instance, reordering methods available include P^3 (Hellerman and Rarick, 1971), hierarchical partitioning (Lin and Mah, 1977), and our modifications of these methods. Solution methods available include Gaussian elimination and the methods described by Stadtherr (1979) and Stadtherr and Wood (1980). Work comparing various combinations of reordering and solution methods is in progress. Using a code based on the solution method given by Stadtherr and Wood (1980), we are able to solve systems of several thousand equations without problem decomposition. For larger problems decomposition is required. Although the current version of SEQUEL does not have a decomposition capability, methods have been described by Westerberg and Berna (1978) and Stadtherr and Hilton (1982).

NUMERICAL STUDIES

Problem Set 1

The studies reported here involve the reliability with which convergence is obtained from the so-called "worst-case" initial guess situation described above. We compare the performance in this regard of the three initial guess types given above and of several different linearization strategies. What if any effect the method of handling thermophysical properties has in this regard is also considered in preliminary fashion.

For the studies reported here, we used ten problems involving the three flowsheets shown in Figures 1, 6, 7, all of which are familiar in the flowsheeting literature. As shown in Table 1, there are two flash with recycle problems, four ammonia manufacture problems and four Cavett problems. The problems differ with respect to which variables are specified by the user. Note that problems 5, 6, 8, 9, and 10 have the feed stream and equipment parameters specified and thus can be categorized as simulation problems. The remaining problems are design problems with one or more output flows specified. The problems may also be classified based on the type of flash specification used. Problems 1, 3, 5, 7, 8, and 9 have flash temperatures specified; we refer to this as the isothermal flash case. In other problems, the flash heat loads are specified; we refer to

this as the adiabatic flash case even though the heat load may be specified to be nonzero.

For the flash with recycle problem a 50% recycle rate is used. For the ammonia problem the precent conversion of the limiting component is set at 25%. Also, since this version of SEQUEL has no compressor module, pump and heater modules are specified to provide the

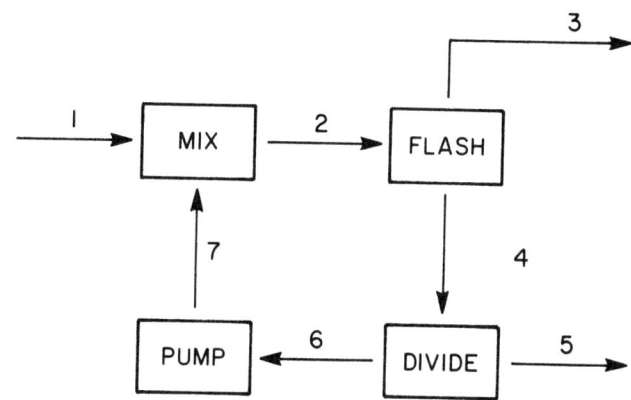

Figure 6. Flowsheet for flash with recycle problem used as example.

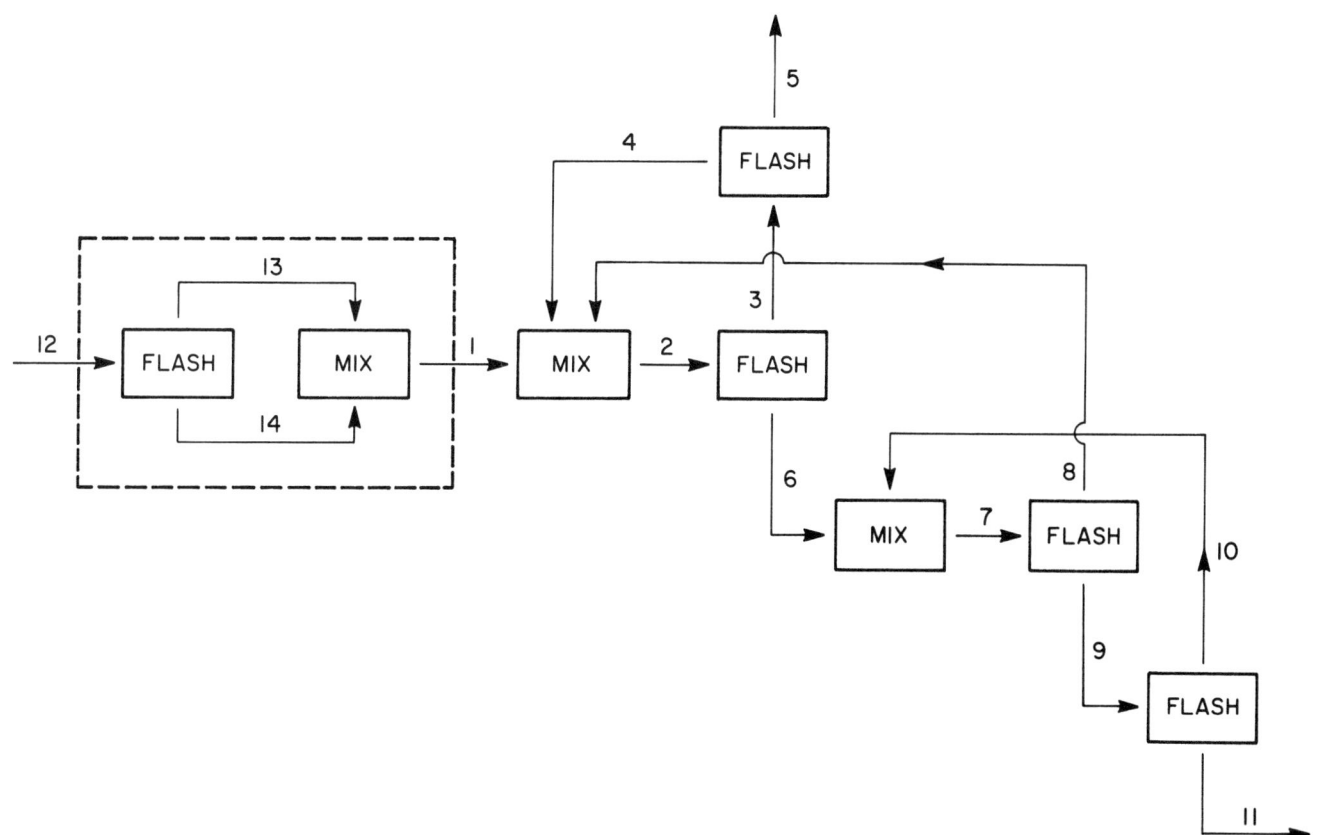

Figure 7. Flowsheet for Cavett problem used as example. Dashed area indicates flash/mix inserted to test condition of feed.

Table 1. Summary of ten test problems used in Problem Set 1.

Flowsheet	Problem Number	Specifications
Flash with Recycle (6 components)	1	n_{11}, T_1, P_1, n_{33}, n_{54}, $n_{31} = 0.994\, n_{11}$, $n_{56} = .69\, F_5$; all unit parameters (T and P for flash)
	2	same as #1, but Q specified for flash, not T
NH_3 Synthesis (5 components)	3	T_1, P_1, $n_{11,1}$, $n_{11,2}$, $n_{16,3}$, $n_{16,4}$, $n_{16,5}$; all unit parameters (T and P for flashes)
	4	Same as #3, but Q specified for flashes, not T
	5	T_1, P_1, $\underline{n}_1$; all unit parameters (T and P for flashes)
	6	same as #5, but Q specified for flashes, not T
Cavett Problem (16 components)	7	T_1, P_1, F_5, F_{11}, $n_{5,1}$; $n_{11,4}$, $n_{11,5}$, ..., $n_{11,16}$; T and P specified for flashes; material balance only
	8	T_1, P_1, $\underline{n}_1$; T and P specified for flashes; material balance only
	9	same as #8, but energy balance also performed
	10	same as #9, but Q specified for flashes, not T

desired temperature and pressure changes. For the Cavett problem a sixteen component system was used. A flash-mixer combination is inserted into the feed stream as a means of testing the feed condition. It should be noted that the largest problem considered here has only 365 equations and thus comes nowhere close to taxing the sparse matrix handling capabilities of SEQUEL. Comparisons of the various sparse matrix routines available in SEQUEL will be presented elsewhere.

The linearization strategies compared are summarized in Table 2. The first strategy is pure Newton-Raphson. The next four are Newton-Raphson with step size relaxation; these differ in how the relaxation parameter r is determined. In strategy #2, r is selected based on a standard norm reduction scheme. Strategies #3-5 employ a scheme that is more primitive, but which performs better on some problems. In these strategies the user supplies an initial value of r and a percentage increment, which is implemented after each iteration in which improvement is observed. The remaining strategies involve a blending of first and second order linearizations, as discussed above. The user supplies an initial blending parameter that indicates the fraction of a second-order linearization used. Thus $\beta = 1$ indicates a Newton-Raphson linearization and $\beta = 0$ a pure first-order linearization. The user also specifies an absolute increment $\Delta\beta$ that is applied either after every iteration or only after iterations in which improvement occurs. In our experience, it is not desirable to remain in the blended ($\beta < 1$) mode for too many iterations, thus in general the values of $\Delta\beta$ used are relatively large.

Table 2. Summary of linearization strategies used in Problem Set 1.

Method	Strategy Number	Parameters	
Newton-Raphson	1		
Newton-Raphson with step size relaxation	2	r set by norm reduction:	
	3	initial r = 0.8,	increment = 10%
	4	= 0.5	= 10%
	5	= 0.2	= 10%
Blended first-order and second-order linearizations	6	Initial = 0,	$\Delta\beta$ = 0.1*
	7	= 0	= 0.1
	8	= 0.5	= 0.5*
	9	= 0.5	= 0.1
	10	= 0.8	= 0.05
	11	= 0.9	= 0.1
	12	= 0	= 1.

*indicates $\Delta\beta$ applied only after iteration during which improvement occurred; otherwise $\Delta\beta$ applied after every iteration.

Since the comparison of the strategies

for handling thermophysical property calculation is not the principal concern of this problem set, somewhat simplified versions of the three strategies discussed above were used. In particular, no dew or bubble point calculations were performed. It is for this reason a flash/mix is inserted in the feed stream to the Cavett problem to check the phase condition. However, the fundamental differences between the three cases are still maintained, and each case presents a distinct problem to the various equation solving routines.

Each of the ten problems was solved using the twelve linearization strategies on the nine possible combinations of initial guess types and thermodynamics cases, the only exception being that the ammonia problems were not solved for thermodynamics Case 3. Thus a total of 936 solutions were attempted. Table 3 summarizes the success rate for the various guess types and thermodynamics cases. The entries in this table indicate what percentage of the twelve linearization strategies were successful on each type of problem. These success rates give a very general indication of how difficult it was to converge a given problem using the twelve linearizations selected. Of course for a different set of linearization strategies, different results might be obtained. Nevertheless, while the results here are not conclusive, they are interesting and in many respects not surprising. First of all, we note that the overall success rate was not particularly good. Since "worst-case" initial guesses were used, this is not too surprising. It does indicate however, the need for better methods of internally generating initial guesses, and the need to consider other approaches, such as the dogleg method, for solving the nonlinear equations. Comparing the initial guess schemes we see that the heuristic approach of Type 3 is in general the best. The development of a better set of heuristics might further improve the success rate.

Comparing thermodynamics Cases 1 and 2 we note that in general Case 2 has a better success rate. This is as expected, since in Case 1 the Jacobian does not include deriatives of K-values with respect to temperature or pressure and thus does not provide an accurate linearization. It is interesting that on the isothermal flash problems (1,3,5,7,8,9), for which the K vs. T information is not important, Case 1 in general seems preferable to Case 2. As expected, it is clear that except in problems for which temperatures are specified (or closely guessed) for all equilibrium stages, it is desirable to linearize the models for K-value prediction and include these in the overall equation set. For Case 3 the results are incomplete, but suggest worse convergence behavior than Case 2. This observation is discussed in more detail below.

Table 4 summarizes the success rate for the various linearization strategies. The entries in this Table indicate how successful each linearization was for solving each problem using all the different combinations of initial guess type and thermodynamics case. For instance, linearization #5 was successful in six of the nine attempted solutions of problem 1. These success rates give a general indication of how reliable each linearization type was under a variety of circumstances. Of course for a different set of initial guesses different results might be obtained. Looking at these overall results, we note that the blended linearizations #6 and #7, which use an initial $\beta = 0$ and a $\Delta\beta = 0.1$, appear the most successful. For most problems, these strategies performed the best or nearly best overall. For some problems a different linearization seems appropriate however. For instance, #9 performed best on problem 7 and #12 best on problem 4. In fact, linearization #9 was the only strategy to solve all the problems at least once. If we focus just on the attempted solutions using guess Type 3, and thermodynamics Case 2 (the best combination from Table 3), linearizations #6 and #7 solve 70 and 80 percent of the problems respectively, but the norm reduction strategy #2 solves 90% of the problems. Thus, it is difficult to make any specific conclusions regarding the different linearizations. We can say in general however, that the blending of first and second order linearizations appears to be a useful approach. While an initial $\beta = 0$ and $\Delta\beta = 0.1$ seem to be relatively good blending parameters, in general it appears that the best blending parameters will vary from problem to problem. This

Table 3. Success rate (%) for different guess types and thermodynamics cases. See text for more detailed information.

Thermodynamics:	Case 1			Case 2			Case 3		
Guess Type:	1	2	3	1	2	3	1	2	3
Problem: 1	92	92	100	92	100	100	92	100	100
2	0	8	17	67	100	100	83	100	92
3	42	17	92	42	17	92	-	-	-
4	17	17	17	33	42	75	-	-	-
5	42	0	67	42	0	67	-	-	-
6	33	0	8	50	0	42	-	-	-
7	0	17	33	25	17	33	0	0	8
8	25	17	67	25	17	33	0	17	25
9	25	17	58	25	17	33	0	17	25
10	0	0	0	25	0	17	0	17	33
Overall	28	18	46	40	30	60	29	42	47

Table 4. Success rate (%) of different linearization strategies on the test problems. See text for more detailed explanation.

Problem Type:	DESIGN					SIMULATION					OVERALL
Flash:	Isothermal			Adiabatic		Isothermal			Adiabatic		
Prob. #:	1	3	7	2	4	8	5	9	6	10	
Linearization:											
1	100	0	11	67	0	0	33	0	17	0	24
2	100	33	11	67	17	11	33	11	17	0	31
3	100	33	0	67	33	11	33	0	17	0	29
4	100	33	11	56	33	11	33	11	33	0	32
5	67	67	22	67	17	11	33	11	17	0	31
6	100	100	0	67	50	78	66	78	33	44	62
7	89	100	0	67	67	78	66	78	50	44	63
8	100	67	0	67	33	22	33	33	17	11	38
9	100	33	33	67	33	56	33	44	33	22	47
10	100	33	11	56	17	22	0	22	0	0	28
11	100	33	22	56	17	0	33	0	0	0	27
12	100	67	22	56	83	0	33	0	33	0	37

suggests that, as noted in Gorczynski et al. (1979), it may be desirable to use different blending parameters for different parts of the process, rather than one set of parameters for the overall flowsheet. The "goodness" of the initial guess is another factor that must be considered in choosing the blending parameters. In general, for better initial guesses than those used here, a wider range of parameters will be successful and numerical efficiency will govern the choice of β and $\Delta\beta$, since it will be desirable to move more quickly into the pure second-order linearization. What is needed is a systematic procedure for selecting an appropriate set of blending parameters for a given situation.

Though the studies presented here focus on the reliability with which convergence is obtained from a poor initial guess, it is also appropriate to comment on numerical efficiency. For the problems solved here, the solution times and number of iterations vary with the guess type, linearization strategy, etc. For the flash with recycle problem a typical solution time is about 0.1 second and about seven iterations are needed. For the NH_3 problem, typical figures are 0.4 seconds and 15 iterations. For the Cavett problem, about sixteen to twenty iterations are needed and times range from about one second for the isothermal cases to about three seconds for the adiabatic case.

These computational times were obtained on a CDC Cyber 175 computer using the FTN compiler under OPT = 1. The number of iterations required reflects the fact that blending in a first order linearization slows the rate of convergence. For this reason, as noted above, when better initial guesses are available, one should avoid staying in the blended mode for many iterations.

Problem Set 2

In this set of problems we look at the convergence behavior from a set of "best-case" initial guesses. These guesses were generated by taking the known solution and randomly perturbing the values in it a minimum of 5% and a maximum of 10, 15, 20, 25, or 30%. Thus five levels of initial guess accuracy were considered. For each problem, three different initial guesses were generated for each level of accuracy.

The problems solved were based on the ammonia and Cavett flowsheets discussed above. The thermophysical properties were handled using either Case 2 or Case 3, thus generating four problems as shown in Table 5. For further details on these and other problems solved see Hilton (1982). For this problem set, dew and bubble point determinations are included using either the Case 2 or Case 3

Table 5. Summary of test problems (Problem set 2).

Problem Number	Flowsheet	Specifications
1	NH$_3$ Synthesis; 5 components; Thermo. Case 2; # of variables = 165	$n_{1,1} - n_{1,5}$, T_1, P_1; T_4, P_4 ("compressor" output); T_6, T_7, T_8; T_{14}, P_{14} ("compressor" output); P and Q of both flashes; $n_{5,2} = 0.1 n_{5,4}$ (this fixes the inerts into the reactor, etc. to be a fixed percentage of the active components)
2	As in 1 but Thermo. Case 3; # of variables = 234	As in 1.
3	Cavett Problem; 12 components; Thermo. Case 2; # of variables = 228	$n_{1,1} - n_{1,12}$, T_1, P_1, H_1; P and Q of all flashes
4	As in 3 but Thermo. Case 3; # of variables = 506	As in 3.

strategy. Thus, the largest problem solved now has 506 equations, which again does not tax the sparse matrix capabilities of the system. The linearization strategies used are the same as for the first set of problems, except that the damped Newton-Raphson strategies 2 and 3 use somewhat more sophisticated norm reduction schemes, as detailed in Table 6. Each of the four problems was solved using the twelve linearization strategies on each of the eighteen different initial guesses. Thus a total of 864 solutions were attempted.

Table 7a summarizes the success rate for the various "best-case" guess types and the two thermodynamics cases. The entries in this table indicate how many solutions out of a possible six were successful. Table 7b compares the overall performance of the Newton-Raphson linearizations (1-5) to the hybrid linearizations (6-12) for the five guess types. Two trends are apparent. Again we note that the Case 2 strategy for computing the thermophysical properties seems more reliable than Case 3. It is not clear why this should be so. Perhaps it is due to the fact that for Case 3 the thermophysical property models are not converged until the end, while for Case 2 they are converged after every iteration. However, this is not a particularly satisfactory explanation. A more probable explanation is that Case 2 is more reliable because the thermophysical property subroutines can employ special-purpose solution algorithms known to be fairly robust for a particular problem, while for Case 3 a general-purpose solution algorithm is applied to the entire equation set. The second trend to note is that the Newton-Raphson-based strategies (1-5) in general perform better than the blended first-second order linearization (6-12). This would seem to indicate that since we are starting near

Table 6. Summary of linearization strategy changes (problem set 2).

Method: Newton-Raphson with step size relaxation

Strategy Number:	Evaluation of relaxation parameter r:
2	$r = \dfrac{(1 + 6n)^{0.5} - 1}{3n}$ where $n = \underline{f}_i^2 / \underline{f}_0^2$ and if $r < 0.05$, $r = 0.05$
3	$r_i = \dfrac{2\underline{f}_{i-1}^2}{2(\underline{f}_i^2 + \underline{f}_{i-1}^2)}$ where if $r < 0.1$, $r = 0.1$ If the norm of f is not reduced: $r_{i+1} = \dfrac{2r_i^2 \underline{f}_{i-1}^2}{2(\underline{f}_i^2 + \underline{f}_{i-1}^2(2r_i - 1))}$ where if $r_{i+1} < 0.1\, r_i$, $r_{i+1} = 0.1 r_i$
2-5	When norm of $\underline{f} < 10$, $r = 1$.

Table 7
a. Success rate (convergences out of 6) of different linearization strategies (problem set 2).
b. Success rate (%) for Newton-Raphson type linearizations and hybrid type linearizations.

(a)

Guess Accuracy:	10%		15%		20%		25%		30%		Overall
Thermo. Case:	2	3	2	3	2	3	2	3	2	3	Success (%)
Linearization: 1	6	4	6	4	6	2	4	1	3	2	63
2	6	6	6	6	5	5	5	3	5	5	87
3	6	5	6	5	6	5	4	4	4	3	80
4	6	5	6	3	6	3	3	2	6	2	70
5	6	6	6	5	5	4	5	2	4	2	75
6	5	3	3	4	3	3	3	2	3	2	52
7	4	3	3	3	2	3	2	3	3	2	47
8	5	5	6	3	4	1	3	0	5	4	60
9	4	4	6	4	5	2	2	2	4	2	58
10	5	5	6	3	5	2	4	3	4	3	67
11	6	5	6	3	4	2	3	1	3	4	62
12	3	4	2	1	1	0	3	1	2	0	28
Total(%)	86	76	86	61	72	44	57	33	64	43	

(b)

Guess Accuracy:	10%		15%		20%		25%		30%	
Thermo. Case:	2	3	2	3	2	3	2	3	2	3
Linearization Type:										
Newton-Raphson (Overall % Success for linearizations 1-5)	100	87	100	77	93	63	70	40	73	47
Hybrid (Overall % Success for linearizations 6-12)	76	69	76	50	57	31	48	29	57	40

the solution, the more accurate second-order linearization should be used from the start. Overall the damped Newton-Raphson strategies 2 and 3 perform quite well. However there is still room for improvements in robustness even from relatively good initializations. We believe the dogleg approach (Powell, 1970; Chen and Stadtherr, 1981) to be particularly promising in this regard.

CONCLUDING REMARKS

A new equation-based flowsheeting system, SEQUEL, is being developed. The system uses a simple executive routine that efficiently generates the equations to be solved, and is interfaced to a set of powerful sparse matrix routines capable of solving several thousand equations without problem decomposition. For larger problems decomposition may be required. Though SEQUEL does not currently have a decomposition capability, the techniques given by Westerberg and Berna (1978) and Stadtherr and Hilton (1982) may be used in this context.

SEQUEL is designed for use in studying

various computational strategies for equation-based flowsheeting. The studies discussed here indicate that simple heuristics can be developed to improve internally generated initial guesses when the user supplies no such information externally. Also the blending of first and second order linearizations appears to be a useful approach to improve convergence from poor guesses. However, choosing the best set of blending parameters is not straightforward. For good initial guesses, as well as some poor guesses, a damped Newton-Raphson approach appears quite effective.

ACKNOWLEDGEMENT

This work has been supported by the National Science Foundation under Grant CPE 80-12428.

LITERATURE CITED

Bending, M. J. and H. P. Hutchison, "The Calculation of Steady State Incompressible Flow in Large Networks of Pipes," Chem. Eng. Sci.. 28, 1857 (1973).

Benjamin, D. R., M. H. Locke, and A. W. Westerberg, "Interactive Programs for Process Design," pres. at AIChE Meeting, Detroit, August 1981.

Book, N. L. and W. F. Ramirez, "The Structural Analysis and Solution of Systems of Algebraic Design Equations," presented at AIChE National Meeting, Philadelphia, June 1978.

Chen, H. S. and M. A. Stadtherr, "A Modification of Powell's Dogleg Method for Solving Systems of Nonlinear Equations," Comput. Chem. Eng., 5, 143 (1981).

Coup, T. G., E. H. Chimowitz, A. Blonz and L. F. Stutzman, "An Equation Analyzer Package for the Manipulation of Mathematical Expressions - I," Comput. Chem. Eng., 5, 151 (1981).

Evans, L. B., "Advances in Process Flowsheeting Systems," in Foundations of Computer-Aided Chemical Process Design, Vol. 1, (ed. R. S. H. Mah and W. D. Seider), Engineering Foundation, New York (1981).

Gorczynski, E. W. and H. P. Hutchison, "Towards a Quasilinear Process Simulator: I. Fundamental Ideas," Comput. Chem. Eng. 2, 189 (1978).

Gorczynski, E. W., H. P. Hutchison, and A. R. M. Wajih, "Development of a Modularly Organized Equation-Oriented Process Simulator," Comput. Chem. Eng. 3, 353 (1979).

Hanyak, M. E., "Textual Expansion of Chemical Process Flowsheets into Algebraic Equation Sets," Comput. Chem. Eng. 4, 223 (1980).

Hellerman, E. and D. Rarick, "Reinversion with the Preassigned Pivot Sequence," Math. Programming, 1, 195 (1971).

Hilton, C. M., "Numerical Studies in Equation-Based Chemical Process Flowsheeting," Ph.D. Thesis, University of Illinois (1982).

Hlavacek, V. "Analysis of a Complex Plant - Steady State and Transient Behavior," Comput. Chem. Eng., 1, 75 (1977).

Hutchison, H. P. and C. F. Shewchuk, "Computational Method for Multiple Distillation Towers," Trans. I. Chem. Eng., 52, 325 (1974).

Kubicek, M., V. Hlavacek, and F. Prochaska, "Global Modular Newton-Raphson Technique for Simulation of an Interconnected Plant Applied to Complex Rectification Columns," Chem. Eng. Sci., 31, 277 (1976).

Leigh, M. J., G. D. D. Jackson, R. W. H. Sargent, "SPEED-UP - A Computer-Based System for the Design of Chemical Processes," presented at CAD-74, Imperial College, London, Sept. 24-27, 1974.

Lin, T. D. and R. S. H. Mah, "A Sparse Computation System for Process Design and Simulation: I. Data Structures and Processing Techniques," AIChE J., 24, 830 (1978).

Lin, T. D. and R. S. Mah, "Hierarchical Partition - A New Optimal Pivoting Algorithm," Math Programming, 12, 260 (1977).

Mah, R. S. H. "Pipeline Network Calculations Using Sparse Computation Techniques," Chem. Eng. Sci., 29, 1629 (1974).

Mah, R. S. H. and T. D. Lin, "A Sparse Computation System for Process Design and Simulation: II. A Performance Evaluation Based on the Simulation of a Natural Gas Liquefaction Process," AIChE J., 24, 839 (1978).

Motard, R. L., M. Sacham, and E. M. Rosen, "Steady State Chemical Process Simulation," AIChE J., 21, 417 (1975).

Powell, M. J. D., "A Hybrid Method for Nonlinear Equations," in *Numerical Methods for Nonlinear Algebraic Equations*, (ed. P. Rabinowitz), Gordon & Breach, New York (1970).

Reid, R. C., J. M. Prausnitz, and T. K. Sherwood, *The Properties of Gases and Liquids*, 3rd ed., McGraw-Hill, New York (1977).

Rosen, E. M., "Steady State Chemical Process Simulation - A State-Of-The Art Review," in *Computer Applications to Chemical Engineering*, ACS Symp. Ser. No. 124 (ed. G. V. Reklaitis and R. G. Squires), ACS, Washington, D.C. (1980).

Stadtherr, M. A., "Maintaining Sparsity in Process Design Calculations," *AIChE J.* 25, 609 (1979).

Stadtherr, M. A. and C. M. Hilton, "On Efficient Solution of Large-Scale Newton-Raphson-Based Flowsheeting Problems in Limited Core," *Comput. Chem. Eng.*, in press (1982).

Stadtherr, M. A. and E. S. Wood, "Exploiting Border Structure in Solving Large Sparse Linear Systems in Bordered Triangular Form," *Comput. Chem. Eng.* 4, (1980).

Westerberg, A. W. and T. J. Berna, "Decomposition of Very Large-Scale Newton-Raphson Based Flowsheeting Problems," *Comput. Chem. Eng.* 2, 61 (1978).

Westerberg, A. W. and S. W. Director, "A Modified Least Squares Algorithm for Solving Sparse n x n Sets of Nonlinear Equations," *Comput. Chem. Eng.* 2, 77 (1978).

Westerberg, A. W., H. P. Hutchison, R. L. Motard and P. Winter, *Process Flowsheeting*, Cambridge University Press, Cambridge (1979).

AN EQUATION ORIENTED APPROACH TO THE STRUCTURING AND SOLUTION OF CHEMICAL PROCESS DESIGN PROBLEMS

A new method of representing and storing a system of equations has been developed for the equation oriented approach to the structuring and solution of design problems. The representation of the equations is powerful and flexible in that the structuring of the calculations and the computation of the solutions are efficiently performed without symbolic manipulation of the equations.

MARILYN J. KOLBET MATTIONE
W. JEFF MEIER
NEIL L. BOOK
Department of Chemical Engineering
University of Missouri-Rolla
Rolla, MO 65401

Process design, analysis, and simulation require the solution of systems of equations that model the performance of process equipment. Mathematically, the objective is to obtain feasible solutions for the systems of equations as efficiently as possible. Mathematical models of processes are often sparse and contain numerous equations, many of which are nonlinear. Further, there are generally more variables than independent equations in process design studies, hence design variables must have numerical values assigned to them before the equations can be solved. The values of many of the design variables, such as feed flow rates, composition, temperature, pressure, and product specifications, are specified by the design problem, but the values of others are either fixed by the designer from historical data on equipment performance or are optimized. However, many different design problems can be developed from a mathematical model by changing either the selected variables or the numerical values of the variables. Thus, where a designer has some freedom in selecting the design variables, the selection should yield a system of equations that can be efficiently solved. The optimization procedures, which solve modeling equations numerous times, then would obtain increased efficiency for each solution.

Mathematical models can be developed so that all the equations are not independent. A set of redundant equations must be removed so that all the remaining equations are independent before a solution is obtained. The selection of redundant equations should also be made to yield an efficient solution.

Once the design variables have been selected and assigned values and the redundant equations removed, the designer should order the equations to represent an efficient solution and then compute the solution. To solve a system of equations efficiently, a designer may wish to combine acyclic and cyclic solution methods. In acyclic or one-at-a-time solutions, a single equation is solved for a single variable, and in cyclic solutions, a set of equations is solved for a set of variables iteratively or simultaneously. The equations to be solved in acyclic or cyclic solutions may be linear or nonlinear.

An equation oriented approach is a method which combines the structuring of the calculations and the solution of the equations. Each equation is analyzed and

---

M.J.K. Mattione is now with Washington University, St. Louis, MO.

W.J. Meier is now with Conoco, Inc., Ponca City, OK.

0065-8812-82-5809-0214-$2.00.
© The American Institute of Chemical Engineers, 1982

manipulated to determine and compute the solution to the system of equations. This is in constrast to a modular approach in which groups of ordered equations are manipulated. The many different design problems which can be created from a single set of modeling equations, make it necessary for a designer to use an equation oriented approach to obtain a solution efficiently.

THE FUNCTIONALITY MATRIX

The structure of a system of algebraic equations (the sparsity, linearity, and nonlinearity) is conveniently expressed by the functionality matrix. Each variable in each equation is required to fit one of the functional forms given in Table 1. The functionality matrix is created by entering the functional form designation at the intersection of the column representing the variable and the row representing the equation of an occurrence matrix. Linear variables are easily identified. Bilinear and trilinear forms, which can degenerate into linear forms, are also identified. An example of the functionality matrix for a system of equations is given in Figure 1.

The functionality matrix is an explicit representation of the structure of the system of equations. The equations may be reproduced from the functionality matrix except for the values of the numerical constants. For example, Equation (3) is:

$$c_1 \ln(x_3) + c_2 x_5^{c_3} + c_4 = 0$$

in which c_1, c_2, c_3, and c_4 are numerical constants, which cannot be determined from the functionality matrix alone.

Provisions have been made for the storage of two characters in each of the matrix elements so that the variables that form the products in bilinear and trilinear forms can be identified. The second character is necessary when more than one bilinear or trilinear form exists in a single equation.

EQUATION STORAGE

The functionality matrix specifies a system of nonlinear algebraic equations to within a set of numerical constants. A system of equations can be completely described by the functionality matrix; two coefficient matrices that contain the constants C_i and C_j, and a vector that contains a constant for each equation, as shown in Figure 2. Because both the functionality matrix and the coefficient matrices are often large and sparse, it is convenient and efficient to store the matrices as a row oriented, column linked list (Lin and Mah, 1978a). Five sets of vectors are necessary. There is a set of vectors associated with 1) the system variables, 2) the columns of the matrices, 3) the system equations, 4) the rows of the matrices, and 5) the entries of the matrices.

An example of the list storage is shown in Figure 3. The variable and equation vectors indicate the column and row in which a variable or equation is stored. Note that the equations and variables may be arbitrarily assigned numbers, and these need not be consecutive. The column and row vectors contain a cross reference to the variables and equations stored in them. The column and row pointer vectors indicate the entry number of the first entry in the column and row. The column and row degrees of freedom provide the number of nonzero entries in the column and row.

The entry lists contain the functional form designation and the numerical constants, C_i and C_j, associated with each matrix entry, and the row and column of the entry are cross referenced. The row oriented, column linked list structure stores row entries in consecutive locations. The first entry in the row can be easily located by using the row pointer. The column link indicates the entry number of the next entry in the same column of the matrix. The first column entry is contained in the column pointer.

Although there are several lists required to store a system of equations, many of the lists contain small integer values, which can be compactly stored. Sparse matrices are efficiently stored, because no storage locations are required for the empty matrix entries.

The system of equations may be exactly reproduced from the five sets of lists. Therefore, the lists completely represent the system of equations. Further, operations required to structure calculations—interchanging the entries of the two rows or two columns—and to compute the solution—solve an equation for a variable or determine the value of a constant or function—are both rapidly and easily accomplished with the list structure.

STRUCTURING THE CALCULATIONS

There are two approaches to structuring calculations in sparse, nonlinear systems: simultaneous linearization and decomposition. In decomposition, a large system of equations is decomposed into smaller subsystems such that the subsystems may be solved sequentially. The computational effort required to solve a dense subsystem of equations is roughly proportional to the cubic power of the number of equations in the subsystem. Therefore, the goal of decomposition is the determination of a sequence of subsystems with a small number of equations in each subsystem.

In simultaneous linearization (Lin and Mah, 1978a,b; Gorczynski and Hutchison, 1978; Stadtherr, 1979; Stadtherr and Wood, 1980), the nonlinear system of equations is iteratively linearized and solved through the use of an efficient method for sparse systems of linear simultaneous equations. The calculations are structured by the strategy used in the selection of pivot elements, when the linearized equations are solved. The pivot elements are selected so as to maintain sparsity in the calculations. An initial guess must be supplied for each of the variables in the equations, and this may be a difficult task when the system is large. Hilton and Stadtherr (1981) tested several methods for internally generating initial guesses.

Westerberg and Berna (1978) indicated that the simultaneous linearization approach has an upper limit on the number of equations for which it is effective as a result of the increased effort necessary to determine the best pivot element. A decomposition of the equations is performed within the simultaneous linearization to create a series of smaller problems in which the approach is effective.

Design variables must be selected by the designer in the simultaneous linearization approach developed by Hilton and Stadtherr (1981), whereas the approach developed by Westerberg and Berna (1978) provides the designer with a single set of design variables. This set of design variables may not be acceptable to the designer, but little information is available to indicate the effects of selecting an alternative set of design variables.

Several decomposition methods have been proposed for structuring calculations. The early decomposition algorithms developed by Steward (1962), Steward (1965), and Himmelblau (1967) do not assist in the selection of design variables. Methods for structuring calculations in sequential, modular simulation, such as those proposed by Upadhye and Grens (1976), Pho and Lapidus (1973), and Barkley and Motard (1972), can be used to structure design calculations, but the design variables must be selected before the algorithms are applied.

Structuring algorithms developed by Lee et al. (1966), Christensen and Rudd (1969), Stadtherr et al. (1974), and Ramirez and Vestal (1972) specify a single set of design variables. The algorithm by Book and Ramirez (1978) provides information as to the effects of selecting a large number of sets of design variables. Simultaneously, information about selecting redundant equations is also produced.

These algorithms attempt to minimize the number of tears in cyclic subsystems, thus, they structure the calculations for solution by direct substitution. Lin and Mah (1978a) note that a very long chain of computations typically exists between the guessed tear values and the residuals in the tear equations. This gives rise to sensitivity problems, which may cause divergence even for initial guesses very near the solution. Both simultaneous linearization and decomposition have limitations, therefore, it appears that either a compromise between decomposition and simultaneous linearization must be effected or the properties of a design problem identified which make the problem suitable for one of the approaches. The approach taken by Westerberg and Berna (1978) is to inbed decomposition within a simultaneous linearization approach, whereas the approach taken here is to inbed efficient solution methods within a decomposition approach.

The functionality matrix shown in Figure 4 is an example of one decomposition for the mixer-exchanger-mixer problem of Book and Ramirez (1978). The functionality matrix indicates that the solution may be obtained by solving a set of three nonlinear simultaneous equations followed by a series of four acyclic equations. A set of four linear simultaneous equations is solved next, and the remaining equations are solved acyclically.

The functionality matrix provides the means for identifying systems of linear simultaneous equations. Note that the bilinear form in Equation (27) of Figure 2 is linear when Equations (38), (26), (32), and (27) are solved, because Equation (10) was solved for variable x_{16} earlier in the calculation sequence.

Although the structuring algorithm prepares the cyclic subsystems for solution by direct substitution, it is not necessary that this method be employed. Research is currently being performed to determine the properties of nonlinear, cyclic systems that make them amenable to efficient solution by various methods.

Linear cyclic subsystems, which do not require initial guesses, are readily identified, and the number of variables, which require initial guesses, is generally small compared to the number of system variables. These factors coupled with the detailed information on the selection of design variables are comparative advantages of the decomposition method. However, it remains to be shown whether the simultaneous linearization or decomposition method is more efficient or more robust. The vast increase in efficiency obtained by storing the equations as a row oriented, column linked list makes it possible to structure large systems of equations at modest expense. Extrapolation of data on systems of up to 250 equations indicate that 1000 equations can be structured in approximately one minute, and 10,000 equations in less than 30 minutes of CPU time.

Further improvements in the power of the structuring algorithm can be made by using the information in the functionality matrix. Functional forms, whose letter designations are Q or higher in the alphabet, have solutions that are multiple valued. In typical engineering applications, only one of these solutions is physically meaningful. The structuring algorithm avoids solving for these variables, if possible. Although linear cyclic systems are easily identified in the decomposed system, an effective, rapid method for locating them during the decomposition has not been developed.

COMPUTING THE SOLUTION

The list storage structure and the functionality matrix also provide computational advantages once the design variables are selected and the calculations ordered. When a variable is selected as a design variable, a numerical value is assigned to it. This may be accomplished by placing the assigned value in an additional list associated with the variables that contains the solution. The columns of the functionality matrix containing the design variables are moved to the right, and subsequent structuring of the calculations is performed as if the value of the variable had been substituted into each equation.

When all design variables have been selected, the computation of the solution proceeds in the order indicated in the structured functionality matrix. If the next subsystem to be solved is acyclic, the equation may be solved by determining the function, $f(\underline{x})$, the numerical constants, C_i and C_j, the equation constant, and the functional form designation. Figure 5 depicts the steps that would result in the acyclic solution of the example equations in Figure 1 when variables 2, 5, and 6 are chosen as design variables and assigned the values 1.5, 1, and -2, respectively. If the acyclic solution is multiple valued, then the feasible region may be checked to determine the physically significant root. The coefficient matrix and the constant vector in linear cyclic subsystems are evaluated in a similar manner. When the value of a variable is determined, it is placed in the solution vector.

Initial guesses for the values of variables in nonlinear cyclic subsystems may also be placed in the solution vector. The designer may specify the use of any of the available methods for solving systems of nonlinear simultaneous equations, otherwise the system will default to a particular method. When the solution of the subsystem is determined, the values are placed in the solution vector. In all types of subsystems, the solution is determined without requiring symbolic manipulation of the equations.

An additional computational advantage is obtained when one solves nonlinear cyclic subsystems by using methods that require the evaluation of Jacobian matrices. The acceptable functional forms have simple, known derivatives, therefore, the analytical values of the entries in Jacobian matrices are directly evaluated. An additional entry vector can conveniently store the values of the Jacobian matrix. When the Jacobian is updated in iterative calculations, only the nonlinear elements need be reevaluated. Figure 6 contains the Jacobian matrix for the system of equations evaluated at the solution obtained in Figure 5.

DATA REQUIREMENTS

Currently, the row oriented, column linked lists are developed from data obtained from the system of equations. Figure 7 shows the data for the example problem of Figure 1. There is a data statement for each matrix entry; however, these data are easily generated from the equations.

Some symbolic manipulation of the equations may be required to place the mathematical model of typical chemical processes in an acceptable form. Experience indicates that this is neither difficult nor overly time consuming. However, efforts are being made to expand the list of acceptable functional forms to include those forms most often encountered in design problems, and special emphasis is being placed on integral and differential equations. Ultimately, modules of typical unit processes will be developed to generate the modeling equations internally.

THE EQUATION ORIENTED APPROACH

The functionality matrix combined with the compact list storage of a system of equations provides a basis for efficient equation oriented computations in chemical process design. The equation oriented computations are of broad scope and versatility. Design variables and redundant equations may be selected, acyclic and cyclic computations, whether linear or nonlinear, may be identified and performed without symbolic manipulations, and Jacobian matrices may be analytically evaluated within a single approach.

LITERATURE CITED

Barkley, R. W. and R. L. Motard, "Decomposition of Nets," Chem. Eng. J., 3, 265 (1972).

Book, N. L., and W. F. Ramirez, "The Structural Analysis and Solution of Chemical Process Design Problems," presented at 85th National Meeting, AIChE, Philadelphia (1978).

Christensen, J. H., and D. F. Rudd, "Structuring Design Computations," AIChE J., 15, 94 (1969).

Gorczynski, E. W., and H. P. Hutchison, "Towards a Quasilinear Process Simulator: I. Fundamental Ideas," Comput. Chem. Eng., 2, 189 (1978).

Hilton, C. M., and M. A. Stadtherr, "Development of a New Equation-Based Process Flowsheeting System: Numerical Studies in Equation-Based Process Flowsheeting," presented at the 75th Annual Meeting, AIChE, New Orleans (1981).

Himmelblau, D. M., "Decomposition of Large Scale Systems - II, Systems Containing Nonlinear Elements," Chem. Eng. Sci., 22, 883 (1967).

Lee, W., J. H. Christensen, and D. F. Rudd, "Design Variable Selection to Simplify Process Calculations," AIChE J., 12, 1105 (1966).

Lin, T. D., and R. S. H. Mah, "A Sparse Computation System for Process Design and Simulation: Part I. Data Structures and Processing Techniques," AIChE J., 24, 830 (1978a).

Lin, T. D., and R. S. H. Mah, "A Sparse Computation System for Process Design and Simulation: Part II. A Performance Evaluation Based on the Simulation of a Natural Gas Liquefaction Process," AIChE J., 24, 839 (1978b).

Pho, T. K., and L. Lapidus, "Topics in Computer-Aided Design: Part I. An Optimum Tearing Algorithm for Recycle Systems," AIChE J., 19, 1170 (1973).

Ramirez, W. F., and C. R. Vestal, "Algorithms for Structuring Design Calculations," Chem. Eng. Sci., 27, 2243 (1972).

Stadtherr, M. A., W. A. Gifford, and L. E. Scriven, "Efficient Solution of Sparse Sets of Design Equations," Chem. Eng. Sci., 29, 1025 (1974).

Stadtherr, M. A., "Maintaining Sparsity in Process Design Calculations," AIChE J., 25, 609 (1979).

Stadtherr, M. A., and E. S. Wood, "Exploiting Border Structure in Solving Large Sparse Linear Systems in Bordered Triangular Forms," Comput. Chem. Eng., 4, 191 (1980).

Steward, D. V., "On an Approach to Techniques for the Analysis of the Structure of Large Systems of Equations," SIAM Rev., 4, 321 (1962).

Steward, D. V., "Partitioning and Tearing Systems of Equations," J. SIAM Num. Anal. Ser. B, 2, 345 (1965).

Upadhye, R. S., and E. A. Grens, "Selection of Decompositions for Chemical Process Simulation," AIChE J., 21, 136 (1976).

Westerberg, A. W., and T. J. Berna, "Decomposition of Very Large-Scale Newton-Raphson Based Flowsheeting Problems," Comput. Chem. Eng., 2, 61 (1978).

Table 1. Functional forms of equations.

Form Designation	Functional Form	Equation Form*
A	Linear	$c_i x_i + f(\underline{x}) = 0$
B	Bilinear	$c_i x_i x_j + f(\underline{x}) = 0$
C	Trilinear	$c_i x_i x_j x_k + f(\underline{x}) = 0$
E	Exponential Base e	$c_i e^{x_i} + f(\underline{x}) = 0$
F	Exponential Base 10	$c_i (10)^{x_i} + f(\underline{x}) = 0$
G	Exponential	$c_i c_j^{x_i} + f(\underline{x}) = 0$
H	Hyperbolic Sine	$c_i \sinh(x_i) + f(\underline{x}) = 0$
I	Hyperbolic Cosine	$c_i \cosh(x_i) + f(\underline{x}) = 0$
J	Hyperbolic Tangent	$c_i \tanh(x_i) + f(\underline{x}) = 0$
K	Cubic Form	$c_i x_i^3 + f(\underline{x}) = 0$
L	Natural Logarithm	$c_i \ln(x_i) + f(\underline{x}) = 0$
M	Common Logarithm	$c_i \log(x_i) + f(\underline{x}) = 0$
P	Power Form	$c_i x_i^{c_j} + f(\underline{x}) = 0$
Q	Quadratic Form	$c_i x_i^2 + c_j x_i + f(\underline{x}) = 0$
R	Square Root Form	$c_i \sqrt{x_i} + f(\underline{x}) = 0$
S	Second Order Form	$c_i x_i^2 + f(\underline{x}) = 0$
T	Tangent	$c_i \tan(x_i) + f(\underline{x}) = 0$
U	Sine	$c_i \sin(x_i) + f(\underline{x}) = 0$
V	Cosine	$c_i \cos(x_i) + f(\underline{x}) = 0$

*c_i and c_j are numerical constants and $f(\underline{x})$ is a function which does not contain x_i, x_j, or x_k.

SYSTEM OF EQUATIONS

EQUATION 5 $x_4 + x_6 x_3 - 3 x_2 x_5 = 7$

EQUATION 3 $4 \ln(x_3) - 2(x_5)^{4.5} = 0$

FUNCTIONALITY MATRIX

R\E C→V	1 x_4	2 x_6	3 x_3	4 x_2	5 x_5	F_I
1 5	A	B1	B1	B2	B2	5
2 3			L		P	2
τ_J	1	1	2	1	2	

Figure 1. The functionality matrix of a system of equations.

FUNCTIONALITY MATRIX

R\E C→V	1 4	2 6	3 3	4 2	5 5	F_I
1 5	A	B1	B1	B2	B2	5
2 3			L		P	2
τ_J	1	1	2	1	2	

COEFFICIENT MATRIX C_I

R\E C→V	1 4	2 6	3 3	4 2	5 5
1 5	1	1	1	-3	1
2 3	0	0	4	0	-2

COEFFICIENT MATRIX C_J

R\E C→V	1 4	2 6	3 3	4 2	5 5
1 5	0	0	0	0	0
2 3	0	0	0	0	4.5

EQUATION CONSTANT VECTOR

R	E	C_E
1	5	7
2	3	0

Figure 2. Matrix storage of a system of equations.

VARIABLE VECTORS

VARIABLE NUMBER	COLUMN NUMBER
1	–
2	4
3	3
4	1
5	5
6	2

EQUATION VECTORS

EQUATION NUMBER	ROW NUMBER	EQUATION CONSTANT
1	–	–
2	–	–
3	2	0
4	–	–
5	1	7
6	–	–

COLUMN VECTORS

COLUMN NUMBER	VARIABLE NUMBER	COLUMN POINTER	COLUMN DEGREES OF FREEDOM
1	4	1	1
2	6	2	1
3	3	3	2
4	2	4	1
5	5	5	2

ROW VECTORS

ROW NUMBER	EQUATION NUMBER	ROW POINTER	ROW DEGREES OF FREEDOM
1	5	1	5
2	3	6	2

ENTRY VECTORS

ENTRY NUMBER	FUNCTIONAL FORM DESIGNATION	C_I	C_J	ROW	COLUMN	COLUMN LINK
1	A	1	–	1	1	–
2	B1	1	–	1	2	–
3	B1	1	–	1	3	6
4	B2	-3	–	1	4	–
5	B2	1	–	1	5	7
6	L	4	–	2	3	–
7	P	-2	4.5	2	5	–

Figure 3. List storage of a system of equations.

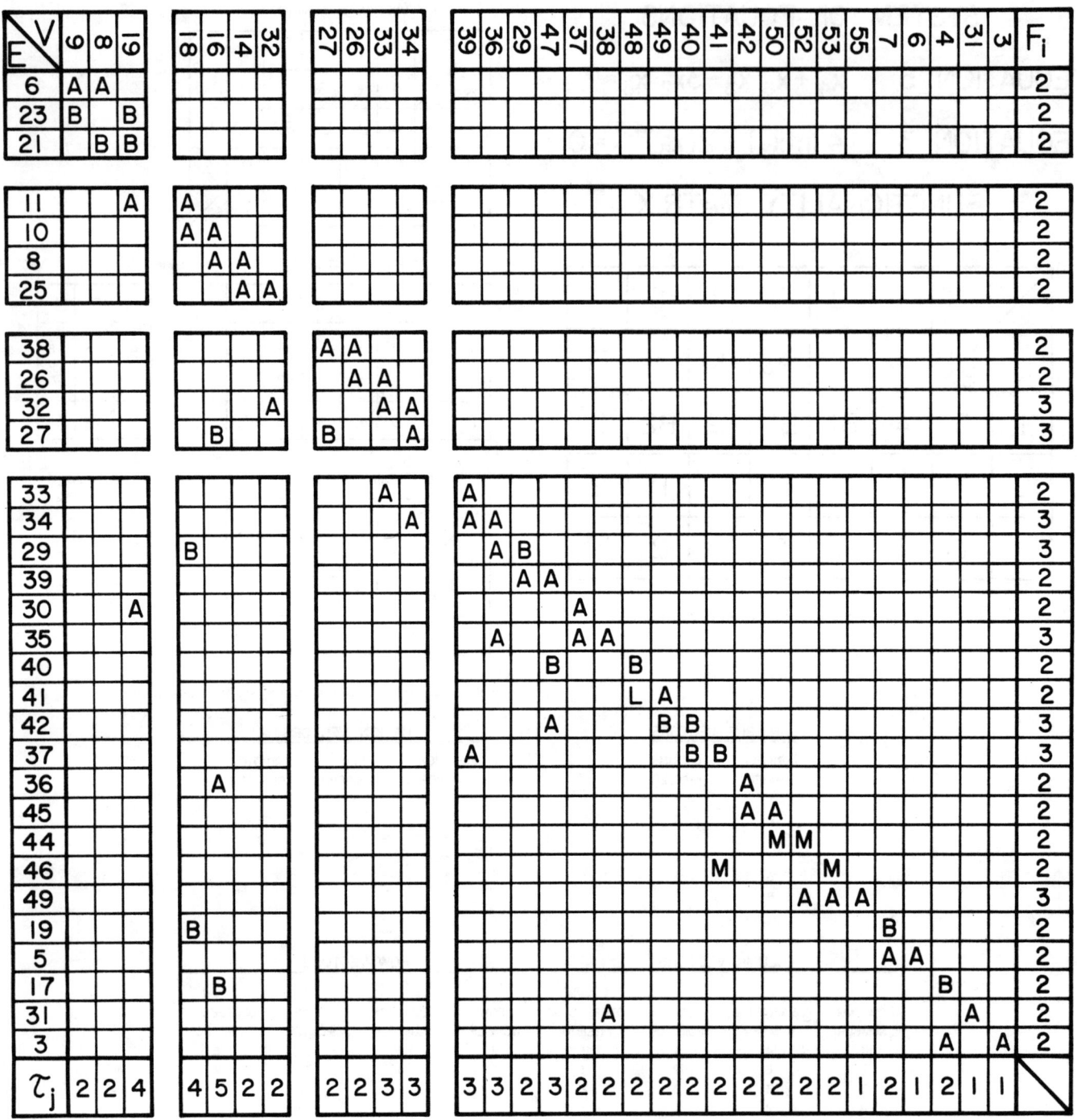

Figure 4. A decomposition for the mixer-exchanger-mixer problem.

VARIABLE NUMBER	COLUMN NUMBER	SOLUTION VECTOR	VARIABLE NUMBER	COLUMN NUMBER	SOLUTION VECTOR	VARIABLE NUMBER	COLUMN NUMBER	SOLUTION VECTOR
1	-	-	1	-	-	1	-	-
2	4	1.5	2	4	1.5	2	4	1.5
3	3	-	3	3	1.65	3	3	1.65
4	1	-	4	1	-	4	1	14.80
5	5	1	5	5	1	5	5	1
6	2	-2	6	2	-2	6	2	-2

(A) DESIGN VARIABLE VALUES ARE PLACED IN SOLUTION VECTOR.

(B) EQUATION 3 IS SOLVED FOR VARIABLE X_3 AND THE VALUE PLACED IN THE SOLUTION VECTOR.

(C) EQUATION 5 IS SOLVED FOR X_4 AND THE SOLUTION COMPLETED.

Figure 5. Acyclic solution of a system of equations.

JACOBIAN MATRIX

R\E\V C	1 x_4	2 x_6	3 x_3	4 x_2	5 x_5
1 5	1	1.65	-2	-3	-4.5
2 3	0	0	2.42	0	-9

LIST STORAGE OF JACOBIAN

ENTRY VECTORS

ENTRY NUMBER	FUNCTIONAL FORM DESIGNATION	C_I	C_J	ROW	COLUMN	COLUMN LINK	JACOBIAN
1	A	1	-	1	1	-	1
2	B1	1	-	1	2	-	1.65
3	B1	1	-	1	3	6	-2
4	B2	-3	-	1	4	-	-3
5	B2	1	-	1	5	7	-4.5
6	L	4	-	2	3	-	2.42
7	P	-2	4.5	2	5	-	-9

Figure 6. Analytical evaluation and storage of a Jacobian matrix.

EQUATION NUMBER	EQUATION CONSTANT	VARIABLE NUMBER	FUNCTIONAL FORM DESIGNATION	C_I	C_J
5	7	4	A	1	-
-	-	6	B1	1	-
-	-	3	B1	1	-
-	-	2	B2	-3	-
-	-	5	B2	1	-
3	0	3	L	4	-
-	-	5	P	-2	4.5

Figure 7. Data requirements for the example problem.

EXPERIENCE WITH ASPEN WHILE SIMULATING A NEW METHANOL PLANT

R.A. KNUDSEN
T. BAILEY
and
L.A. FABIANO

ARCO Chemical Compny
3801 West Chester Pike
Newtown Square, Pennsylvania 19073

ABSTRACT

ASPEN was used to rate the design of a new 2000 ton/day (75,000 kg/hr) methanol plant at various operating conditions. Several levels of complexity in simulation were employed. The present model includes existing ASPEN blocks and two comprehensive user-developed blocks with a third major reactor block planned. The simulation program will provide our engineers with a model capable of predicting equipment performance and limitations and the overall plant performance. Optimum operation will be predicted for any capacity, for alternative feedstocks, for different catalyst ages, and for different equipment fouling. Operator training and plant performance monitoring will be a major application of the simulation.

"World Scale Plant" connotation conveys several important meanings. In the competitive international marketplace, the economics of commodity chemicals production requires large plant size, high plant efficiency and innovative designs. Production cost must be minimized during both the design and operating stages.

High efficiency and enormous size by definition require a large capital outlay. Today operating companies must be more concerned with several factors:

- Marked differences in operations between start and end of run.
- Import or export of substantial energy streams.
- Forced operating at turn-down conditions during slumps in the marketplace.
- Increased operation for production limited operations.
- Unusually high costs associated with start-up and shut-down.

For an appreciation of energy and feed requirements, consider that a 2000 ton per day (75,600 kg/hr) methanol facility consumes approximately 20 billion SCF per year (570 million SCM per year) of natural gas at current cost of $70 million. Capital assets typically range between $200-250 million. This high capital cost is due in part to the high degree of heat and energy recovery. Its design complexity obviously adds to the operating complexity.

It becomes critical that these plants always operate near maximum efficiency. Transitional periods, e.g., less than full capacity, can be very costly if not planned and operated optimally. Flexibility to adapt to alternate feedstocks due to government intervention must also be possible. Catalyst deactivation or "changeovers" and equipment fouling will change the optimum operation with time. How can an operating company prepare to quickly react?

Our proposed solution is to prepare a flowsheet simulation that can predict performance of in-place equipment. This model must be only as regorous as necessary to accurately predict equipment and unit operations. It will require significant manpower but will pay dividends in the end.
The simulation for our methanol plant will serve several purposes. It will:

- Provide a standard calculational technique which is not dependent on changine operations personnel.
- Provide a consistent data base for comparing operations at different times.
- Store specific equipment physical parameters.
- Store normal and extreme operating conditions.
- Monitor and flag normal/abnormal operations.
- Provide a training tool for new engineers.

PLANT DESCRIPTION

The plant being modelled with the ASPEN simulator (Evans, et.al., 1979) is ARCO Chemical's new 2000 ton/day (75,600 kg/hr) Methanol Plant. This facility located just outside of Houston, Texas is scheduled for start-up in late 1982. The primary feedstock to the plant is pipeline natural gas, and the refined methanol product will meet Federal Grade AA standards. The plant was designed and engineered by Davy-McKee Corporation, and employs ICI technology for methanol synthesis and purification.

A simplified block flow diagram of the process is presented in Figure 1. The pipeline natural gas is first compressed to the plant operating pressure. The compressed gas is then desulfurized, and mixed with process steam prior to the reformer. The steam-gas mixture is reacted over a suitable reforming catalyst to produce an effluent rich in hydrogen and carbon oxides:

$$CH_4 + H_2O \longrightarrow 3H_2 + CO$$

$$CO + H_2O \longrightarrow CO_2 + H_2$$

The high temperature process gas exits the reformer at approximately 1600°F (871°C) and is cooled to recover heat to the process. The cooled reformed gas is then compressed to 1500 psig (10,400,000 Pa), and injected into the methanol synthesis loop. Methanol is produced in this section of the plant via the following reactions:

$$CO + 2H_2 \longrightarrow CH_3OH$$

$$CO_2 + 3H_2 \longrightarrow CH_3OH + H_2O$$

A recycle reactor system with a suitable high activity catalyst is employed to maximize carbon oxide conversion to methanol.

The crude methanol from the synthesis loop contains 15 to 20 mole percent water, as well as trace quantities of other contaminants. These contaminants are removed via fractional distillation in the methanol purification section. Final product quality is 99.85 mole percent (minimum) methanol.

The primary utility requirement for the plant is high pressure steam. Steam is utilized by the process in several ways:

- process steam to the reformer
- turbine drives for compressors
- distillation column reboil heat

Essentially all the steam required for the process is generated by recovery of heat from the gases. A complex network of heat exchangers is employed in the design to maximize the recovery of heat from the and flue gases.

BASIC SIMULATION PHILOSOPHY

Our basic approach can be described as a "layered approach". We started by connecting the simplest ASPEN blocks to describe all significant unit operations on the existing Process Flow Diagrams (PFD). More rigorous blocks were used when the old blocks were not sufficiently accurate. This generated the ASPEN flowsheet framework which could easily be upgraded, and provided a check on the ASPEN simulator, particularly the unit operations and thermodynamics. The approach also minimized computer time for any run since the models were only as rigorous as needed. If a "bug" was encountered, we were able to devise an alternate approach to circumvent the problem until it was corrected by the ASPEN staff.

The best simulation employs the simplest blocks which are consistant with the desired overall accuracy. "Full blown" rigorous models consume computer time, may not significantly improve accuracy, and are more difficult to modify to fit actual plant performance.

A "level zero" model tested the basic ASPEN system by using our methanol PFD. Recycles loops were minimized by dividing the design flowsheets into distinct sections. Heat and material balances were checked by specifying flowsheet variables, such as temperature or heat duty, instead of size parameters such as heat transfer area.

Higher level models used equipment size parameters to rate the actual equipment performance. Other upgrades included:

- "steady state process con-trols"
- proprietary high-pressure thermodynamics
- rigorous multistage compressor user developed block
- reformer duct user developed block
- proprietary synthesis re-actor model

An example of the "layered approach" for a heat exchanger is shown in Figures 2 thru 5. The "level zero" model (Figure 2) evaluates one side of the heat exchanger, E-7000S, based on the specified outlet temperature. It calculates an information stream, Q-7000S, leaving E-7000S; and feeds it to the other side of the heat exchanger, E-7000T. E-7000T then determines stream 4 based on this heat transfer rate. This "level zero" model is good for a new plant design where the process engineer can then design a heat exchanger to meet these heat transfer requirements.

For an existing plant design, a rating model based on actual installed equipment would be more useful. The "level one" model, shown in Figure 3, is a rating model. It is exactly the same as the "level zero" model in Figure 2 except that a "DESIGN-SPEC" has been added and labelled Q0. This "DESIGN-SPEC" varies the outlet temperature from E-7000S to make the absolute value of Q-7000S equal to the heat transfer calculated by $UA \Delta T_{LM}f$, based on the sampled variables. For our typical "level one" model, we used an HTRI rigorous design as the basis, and varied the inside and outside heat transfer coefficients according to the mass flow.

The "level two" model is shown in Figure 4 and is the same as the "level one" model except that a pressure drop correlation is used. Two "FORTRAN" blocks, DPOS and DPOT, set the pressure drops for E-7000S and E-7000T, respectively, by noting the inlet mass flowrate and density. The pressure drop is varied according to the change in velocity head, i.e., it is proportional to M^2/ρ where M is the mass flowrate and ρ is the density.

The "level three" model is shown in Figure 5. It is helpful for a pinch point which can occur when consolidation or evaporation is present. It is essentially the same as "level two" except that the exchanger is evaluated twice. A "DUPL" block repeats material streams; two new user added blocks then duplicate, split the heat exchange duty into equal heat transfer increments and change sign of the heat stream. Note q00S1 = q00T1 and q00S2 = q00T2 condensing or vaporizing heat transfer coefficients are combined with sensible heat transfer coefficients according to the HTRI equations. The overall heat transfer rate is determined via a "DESIGN-SPEC" on the temperature leaving E-7000S and is based on the sum of $U_i A_i \Delta T_i$ across each heat transfer increment. A calculated temperature crossover would show as a negative $U_i A_i \Delta T_i$, and thus the "DESIGN-SPEC" would seek a new E-7000S outlet temperature. Figure 5 shows two heat transfer increments. More increments can be used if necessary.

For units with distinct sensible heat regions and condensing and/or evaporation regions, a more complicated model is required which involves the bubble point and dew point of the individual stream. We have used this "level four" model, but are not showing it due to the complicated diagram required.

ASPEN has a two-sided heat exchanger model, HEATX, which can rate a heat exchanger based on a specified area. This model permits different heat transfer coefficients for different phase conditions within the exchanger. It was not used because of a "bug" in the HEATX program. However, it can be considered for future models since this "bug" has been corrected.

APPLICATION OF ASPEN

Reforming Section

The reforming section of a modern world scale methanol plant is quite complex in design, and consumes large quantities of energy. The natural gas feed and fuel requirements for a 2000 ton/day (75,600 kg/hr) methanol plant approach 600MM SCFD. Obviously, efficient operation of the reforming section of the plant is essential for minimizing plant operating cost. A model capable of estimating the performance of the actual installed equipment is a valuable tool for optimizing unit operation.

Equipment Description. A simplified schematic of our methanol plant reformer is given in Figure 6. The reforming reactions occur in several catalyst filled tubes which are arranged in the reformer radiant section. These reactions are endothermic and require a rather high operating temperature for maximum methane conversion. The heat required for the process is supplied by a number of burners located in the roof of the reformer radiant box. Fuel to these burners is natural gas supplemented with a hydrogen rich purge gas from the methanol synthesis loop. The hot reformed gas leaving the reformer is cooled to near ambient conditions in a series of exchangers. Almost all the heat is recovered by generating high pressure steam or direct exchange with process streams.

The hot flue gases exiting the reformer radiant box are directed over a series of heat absorbing coils in a convection duct.

These coils preheat the natural gas feed and generate and superheat high pressure steam. A small auxiliary firing chamber is located upstream of the convection duct to control the rate of steam production.

The last coil in the convection duct is a combustion air preheater which preheats the air to approximately 700°F (371°C). Air preheat improves radiant box efficiency, and therefore minimizes furnace fuel requirements.

Equipment Simulation. To simulate the operation of the reforming section of the plant, it was first necessary to develop performance rating blocks for each individual process step in the reformer. The development of these blocks is outlined below.

Radiant Box - Process Side. The steam-methane reforming and water-gas shift reactions occurring in the reformer tubes were simulated with the "RGIBBS" equilibrium reactor block. The feedstock inlet conditions and outlet temperature are specified in the user input, along with the anticipated approach to equilibrium for each of the reactions. The ASPEN SYSOP3 thermodynamics (Soave-Redlich-Kwong) were used in this simulation.

The results from this simulation showed excellent agreement with design data. Calculated residual methane leak was within 0.05 percent (absolute) of design, and the required heat duty was essentially identical to design. Subsequent testing over a wide range of exit temperatures, pressures, and steam/carbon ratios showed similar excellent agreement with published data on equilibrium constants and reaction heat.

The performance of the basic "RGIBBS" block was subsequently enhanced by employing a "FORTRAN" block to calculate the gas pressure drop through the catalyst packed tubes. It is planned to add another FORTRAN block to calculate the temperature approach to equilibrium for each reaction as a function of reformer flowrate and exit temperature.

These values will be used to reset the user supplied default values for approach to equilibrium. (See Figure 7).

Radiant Box - Fireside.

The ASPEN process simulator does not presently contain any unit operation block capable of rating the performance of a fired heater. Since the fuel consumed in the reformer is a major energy stream in the process, it was decided to develop a suitable rating algorithm for the fireside of the reformer.

A proprietary computer model capable of performing a rigorous analysis of top fired reforming furnaces was employed in this work. Numerous simulations were performed while varying the key operating parameters in the reformer radiant section. Analysis of these data lead to the development of a semi-empirical model capable of predicting the fuel and air requirements of the reformer, over a wide range of operating conditions. This simplified model proved easy to tie into the ASPEN framework, and gave results essentially equal to the more rigorous model. Key operating parameters included in the firebox efficiency calculation include the following:

- reformer feedrate
- steam to carbon ratio
- reformer exit temperature
- air preheat temperature
- percent excess air
- process side heat duty
- heat losses to atmosphere
- fuel preheat temperature

Application of this efficiency algorithm to rate reformer performance is illustrated in Figure 8. Required data from the stream vectors for reformer feed and effluent streams (material plus heat streams) are supplied to "FORTRAN" block FUEL. Additional data on the combustion air preheat temperature and percent excess air are also input to this block. This block utilizes these data to calculate the required flowrate and composition of fuel to the reformer.

The required flowrate of combustion air to maintain the desired percent excess air is also calculated. This information is sent to an ASPEN "RSTOICH" block which simulates the combustion of the fuel/air mixture. The flue gas outlet temperature from the "RSTOICH" block is controlled by a "DESIGN SPEC" which equates the heat removed from the hot flue gas to the heat absorbed by the process gas (heat stream QREFM) plus wall losses.

Convection Duct.

The convection duct of the reforming furnace is employed to recover heat from the hot flue gases exiting the radiant box. These gases typically exit the radiant box at temperatures of 1750 to 1850°F (954 to 1010°C). At these temperatures, heat transfer to the heat absorbing tubes via radiation is quite significant. A model such as the ASPEN "HEATX" block is not capable of adequately rating convection coil performance over a range of operating conditions. For this reason, it was decided to develop a suitable user block.

The result of this work was the ASPEN user block "DUCT" which performs detailed analysis of both radiation and convective heat transfer to each coil in the duct. This model calculates the total heat transfer to each tube coil and the resultant changes in flue gas and process fluid temperatures. A total of four inlet and four outlet material streams are evaluated in the block. Values of the heat absorbed in each coil are stored as elements in the REAL matrix used by "DUCT." These values are subsequently accessed by our utility balance block to calculate the total steam generation within the plant. "DUCT" also makes use of the ASPEN feature permitting different thermodynamic systems to be employed within one process block. All steam calculations are performed with SYSOP12 (1967 ASME Steam Tables), while flue gas and process gas properties were evaluated with SYSOP3 (SRK).

Combustion Air Preheater.

Our plant employs a single large rotary type heater to preheat combustion air to the radiant box and auxiliary burners.

Simulation of this equipment is quite similar to the conventional shell and tube exchanger algorithm previously outlined (See BASIC SIMULATION PHILOSOPHY). In this case a "DESIGN-SPEC" was employed to vary the outlet temperature of the combustion air until the following relationship was achieved:

$$M_{AIR} \Delta H_{AIR} = M_{FG} C_{PFG} \Delta T_{FG}$$
$$= U_O A_O \Delta T_{LM}$$

The apparent overall heat transfer coefficient was evaluated as a function of flue gas and air flowrates based on data supplied by the equipment vendor. Combustion air leakage to the flue gas was calculated as a function of differential pressure.

Reformed Gas Heat Recovery Train. The heat exchange train for the hot reformed gas was simulated using the scheme illustrated in the BASIC SIMULATION PHILOSOPHY section. Sensible heat and condensing type exchangers were modeled using the "level two" and "level three" algorithms as required. The design individual heat transfer coefficients were taken from HTRI simulations, and varied within the model as a function of stream flowrate. Plans for future enhancement of these blocks include addition of stream transport properties to account for the impact of composition on heat transfer rates.

Overall Simulation Scheme. A simplified flowsheet of the overall reformer simulation is presented in Figure 9. Note "DESIGN-SPEC" BURN and heat stream QBURN from Figure 8 were omitted to simplify the drawing. The simulation scheme is fairly straight forward, employing a unit operation sequence analogous to actual equipment arrangement. Iterative loops are established to solve for flowsheet variable such as:

- process gas inlet temperature to reformer
- combustion air temperature
- high pressure steam generation in plant

Several other computational loops are used in the complete simulation of the reforming section to iteratively solve for variables such as natural gas compressor speed, desulfurization temperature, reformer outlet pressure, etc. Model convergence has been excellent even with relatively poor initial guesses for unit operation block input. Total computational time for the reforming section simulation is in the order of 80 to 90 seconds on an IBM 370/3033 system.

Compression Section

The reformed gas compression section consists of a large multistage centrifugal compressor driven by a direct coupled steam turbine. A simplified schematic of this equipment is presented in Figure 10. The total horsepower requirement of this machine is typically 20,000 to 40,000 hp (15,000,000 to 30,000,000 W) for a world scale methanol plant. A large percentage of the total plant steam requirement is consumed in the combination extraction/condensing turbine which drives this compressor.

Since the plant compressors are such a large energy consumer, a simulation tool was needed to accurately predict their performance over a wide range of conditions. The standard ASPEN compressor block "COMPR" was found inadequate for use as an equipment rating block. Therefore, it was decided to develop in-house a rigorous compressor rating algorithm. The result of this work was the centrifugal rating model "RIGCOMP." This block performs a rigorous wheel-by-wheel analysis of a centrifugal compressor evaluating such items as gas horsepower, brake horsepower, compressor speed, gas outlet conditions, etc. "RIGCOMP" is general in nature and can predict the performance of any centrifugal compressor. The model has been checked against actual vendor performance curves and shown to be accurate to within approximately 1 percent from surge to choke.

"RIGCOMP" was developed as a stand alone program, and subsequently tied into ASPEN as a user block. This conversion required about two man-weeks, and was accomplished without any major problems. The inclusion of "RIGCOMP" in the plant simulation permits accurate prediction of compressor performance as plant operating conditions vary. In addition, the ASPEN- RIGCOMP combination permits the programmer to more directly simulate actual plant controls. Take for example the system depicted in Figure 11. A pressure downstream of the compressor is controlled at some value by variation of the compressor speed. This system is easily modeled on ASPEN by employing a "DESIGN-SPEC" on downstream pressure which adjusts the speed input to the "RIGCOMP" user block. The simulation algorithm in this case is totally analogous to the actual plant control scheme. The user can independently manipulate the controlled plant variable (i.e. pressure) and study the anticipated impact on compressor operation. The permissible range of plant operation between compressor surge and choke conditions can be simply and directly defined.

The tie-in between the compressor and the steam turbine driver is accomplished by using an ASPEN work information stream. The basic simulation scheme is depicted in Figure 12. The total brake horsepower work stream from the centrifugal compressor is fed to a standard "COMPR" block, which simulates the extraction section of the turbine drive. An initial estimate of the high pressure steam flow is also supplied to the block. This block calculates the outlet steam conditions from the extraction section, as well as the horsepower delivered to the drive shaft. The exit work stream WORK2 represents the additional horsepower required to satisfy compressor demand. This power must be supplied by the condensing section of the turbine, which is also simulated by a "COMPR" block. The medium pressure steam exit TURB1 is first sent to a "FSPLIT" block to simulate the controlled flowrate of extract steam from the turbine.

The remaining steam is input along with stream WORK2 to the condensing turbine TURB2. This block calculates steam outlet conditions as well as horsepower delivered to the drive shaft.

The correct flowrate of high pressure steam to the turbine is obtained when the sum of the horsepower supplied from both sections of the turbine exactly equal compressor needs. In this simulation, the steam flow is controlled via a "DESIGN-SPEC" which manipulates steam flow until the excess work stream WORK3 equals 0.

The exit work stream WORK3 represents the horsepower imbalance between WORK2 and the power delivered by TURB2.

Synthesis - Reactor and Heat Recovery Loop

The principal parts of the methanol synthesis reactor and heat recovery loop are shown in Figure 13. A synthesis converter is fed by a "warm shot" and controlled by a "cold shot". Several shell and tube exchangers provide the converter heat recovery system. A crude methanol separation unit recovers unconverted reactants. Two other principal parts are a recycle compressor and a pressure letdown vessel.

Synthesis Converter.
The synthesis reactor is a multi-staged adiabatic reactor. A four staged synthesis reactor is shown in Figure 14 as a typical example. Sufficient quantity of "cold shot" gas is mixed with each stage reactor effluent to control the feed temperature going to the next stage (Stephens, 1974). The "level zero" methanol converter model used a single "RGIBBS" block to mix and react the "warm shot" and total "cold shot". It then adjusted the outlet temperature with a "DESIGN-SPEC" to make the heat stream be zero since "RGIBBS" is not an adiabatic block. See Figure 15. "RGIBBS" is an equilibrium model. (White and Seider, 1981) It was selected over "REQUIL" because it is a newer and more robust algorithm, and "REQUIL" had several "bugs." "RGIBBS" predicts considerable methane formation if a single approach temperature is specified. Consequently, an approach temperature must be specified for each reaction catalyzed. This converter was fit by setting different temperature approaches to equilibrium for two reactions.

$$CO + 2H_2 \longrightarrow CH_3OH$$

$$CO_2 + H_2 \longrightarrow CO + H_2O$$

Side reactions, such as the formation of dimethyl ether and higher alcohols, were considered by specifying the extent of reaction. Reasonable temperature approaches to equilibrium were fit using this approach, although the results were slightly dependent on the thermodynamic option selected.

"Level one" model (Figure 16) specified a "cold shot" to each of the individual stage reactors, holding the temperature approach to equilibrium constant. Results show the typical temperature "saw-tooth" curve. See Figure 17.

"Level two" model simulates a temperature control for the feed to each reactor stage. See Figure 18. This cannot be done with several "FSPLIT" and "DESIGN-SPEC" blocks. Our ASPEN input used a "DUPL" (duplicate) block and several "MULT" (multiply) blocks with the "DESIGN-SPEC". As with the "level one" model, ASPEN evaluates R-7001 via "RGIBBS" and a "DESIGN-SPEC" QR1 on R-7001 to set R-7001 outlet temperature to make R-7001 adiabatic. The "cold shot" is sent through a "HEATER" block to permit subsequent readjustment of the "cold shot" temperature. The resultant "cold shot" goes to a "DUPL" block from which its product streams go through "MULT" blocks. This creates matter. However, it is used only as a means to an end. Each "MULT" factor is adjusted by a "DESIGN-SPEC" to achieve a specified "MIXER" product temperature. Another "DESIGN SPEC" labelled "RXTCT" adjusts the "CSTEMP" block outlet temperature so that the sum of the "MULT" block factors equal to one. Thus, mass balance is preserved and well as the energy balance leaving "CSTEMP".

"Level three" converter Loop Model is the same as "Level two" except that the "RGIBBS" models with their associated adiabatic "DESIGN-SPEC" are replaced by a proprietary reactor kinetic model. See Figure 19.

Converter Heat Recovery System. Hot converter effluent gas must be sufficiently cooled and condensed before processing in the crude methanol separation unit. A good design recovers this energy to minimize fuel requirements. The temperature control of the converter "warm shot" and "cold shot" streams provide an example of the flexibility of ASPEN. The basic heat recovery scheme is shown in Figure 20. Sufficient hot effluent gas is split to E-7001 to heat warmed compressed recycled gas to the desired "warm shot" temperature. The resulting cooled converter effluent gas then merges with another cooled converter effluent stream from E-7002. This stream heats parts of the compressed recycle gas to achieve the "cold shot" temperature when mixed with the bypass. The remaining converter effluent goes to other heat recovery exchangers.

The "level zero" model is shown in Figure 21. Since the actual flowrates and split ratios are specified, it is not important which side of the exchanger is evaluated for this model first - except the direction of the heat stream obviously changes.

The primary purpose of higher level models is to satisfy the reactor control scheme. This scheme is to set the "cold shot" and "warm shot" temperatures for subsequent temperature control of each reactor stage feed temperature. Other exchangers, not discussed, guarantee that the product is sufficiently cooled and condensed before processing in the crude methanol separation unit.

The calculational order is important in the "level one" model, shown in Figure 22. It is best to evaluate the compressed recycled gas loop first since the "warm shot" and "cold shot" temperatures and flowrates are known. An initial guess of the SP1 split is made, and the "cold shot" temperature "DESIGN-SPEC" TCS is converged by varying the SP1 split until the "cold shot" temperature is the specified value(1). E-7001T is evaluated next, although it could be evaluated immediately after converging "DESIGN-SPEC" E-7002. Next an initial guess is made for the SP3 split, and the E-7002 "DESIGN-SPEC" is converged by varying the SP3 split until E-7002 equals UA $\triangle$ $T_{LM}f$ calculated as described in the BASIC SIMULATION PHILOSOPHY Section. The simulation proceeds to the E7001 "DESIGN-SPEC". Here the SP4 split is varied from an initial guess until q_{7001} equals UA $\triangle$ $T_{LM}f$ for E-7001. Finally, MX2 and E-7003 are evaluated to see if the resulting temperatures obey q_{7003} = UA$\triangle$$T_{LM}f$ for E-7003. If they do not, E-7003 reguesses E-7003T temperature and starts over. This procedure is repeated until "DESIGN-SPEC" E-7003 converges.

High pressure vapor-liquid equilibrium data were required to adequately simulate these exchangers. We wrote a proprietary liquid activity users model.

Higher level models are not shown because they would unduly complicate the diagrams. The basic approaches are shown in the BASIC SIMULATION PHILOSOPHY Section.
During the initial development of these ASPEN models, the "DESIGN SPEC" required the limits on the adjustable variable to be constant. This necessitated using unduely

---

(1) An alternate approach is set the SP1 split ratio during the calculation via a "FORTRAN" block for each E-7003T outlet temperature. This feed forward controller requires an enthalpy balance which can be made with the assistance of a "DUPL" block and several "dummy" heat exchanger blocks.

tight limits to prevent the argument of the logarithm from becoming negative when the LMDT was successful. One approach to circumvent this problem was successful except for some cases where it "converged" to an infeasible solution. This method used the hot stream outlet temperature resulting from the "HEATER" block to calculate the same temperature from an $q = UA \Delta T_{LM} f$ equation. This equation was written as:

$$T_{HOC} = T_{CI} + (T_{HI} - T_{CO}) \exp\left\{\frac{UAf}{q} [(T_{HOG} - T_{CI}) - (T_{HI} - T_{CO})]\right\} \quad (1)$$

where:

T = temperature
U = heat transfer coefficient (function of flowrate)
A = heat transfer area
q = ASPEN heat stream
f = factor to convert LMTD to MTD

subscripts:

H = hot
C = cold
O = outlet
I = inlet
G = guess (from ASPEN block)
C = calculated

Several cases were observed which converged to

$$T_{HOG} - T_{CI} = T_{HI} - T_{CO} \quad (2)$$

depending on the initial guess. However substitution of (2) into

$$q = UA \left[\frac{(T_{HOG} - T_{CI}) - (T_{HI} - T_{CO})}{\ln\left(\frac{T_{HOG} - T_{CI}}{T_{HI} - T_{CO}}\right)} \right] f \quad (3)$$

showed this calculated q differed substantially from the ASPEN heat stream. See Figure 23. This was not investigated further at this time due to the development of the ASPEN subroutine "SETLIM" which permits one to change the limits during the calculation.

Crude Methanol Separation Unit Model. The crude methanol separation unit operates at a high pressure while recovering uncoverted synthesis gas from methanol. This synthesis gas is subsequently recycled to the recycle compressor and methanol converter. Recovery depends on the solubility of these gases in a methanol-water mixture. Proprietary solubility data were inserted into the model using three ASPEN blocks, the previously noted high pressure liquid activity user model and an ASPEN "FORTRAN" block. See Figure 24. The feed stream is first flashed in a "FLASH2" block using the high pressure liquid activity model. The flowrate of condensibles, e.g., water and methanol, in the "FLASH2" liquid product stream is preserved in the "SEP" block, DS-7001. All other components are split as dictated by the proprietary data in the "FORTRAN" block SOL.

Closing the Synthesis Reactor and Heat Recovery Loop. The synthesis reactor and heat loop is completed by adding the recycle gas compressor previously described to the model. This system has two recycle loops. One is a material recycle loop, and the other is basically an energy transfer loop. The material recycle loop is the recycle from the crude methanol separation unit. A recycle stream convergence method must be used. The energy transfer loop is essentially the "cold shot" and the "warm shot" streams. Here we have the option of either using a recycle stream convergence method in which both the synthesis converter and the converter heat recovery system can be evaluated several times, or using a "DUPL" block with some dummy "HEATER" blocks to duplicate the "cold shot" and the "warm shot" streams since we know their flowrate, composition and temperature. The second approach requires a "DESIGN-SPEC" on the pressure drops in HDCS and HDWS, shown in Figure 25. Our study showed that it was easier to manually set the pressure drops in HDCS and HDWS; thus eliminating considerable computational time.

Distillation

There are two parts to the distillation simulation. One is the rigorous stand alone simulation in which the design was checked. This was done early in the project using another simulator. We plan to check the results with RADFRC.

The other is for our overall simulation. This will be a simplified model. Since the feed composition to the columns remain essentially constant, we plan to develop utility curves for steam and cooling water as functions of feed rate. These curves will be based on FRI efficiency correlations and rigorous distillation column runs.

Plant Steam Balance

Steam is generated within the plant at approximately 1500 psig (10,400,000 Pa). This high pressure superheated steam is used to drive the main compressors in the plant. Two other steam pressure levels are also employed within the plant. The medium pressure system is fed from turbine extract and supplies process steam for the reformer. This pressure level is also used for turbine drives on various medium sized turbines in the plant. (See Figure 26). The extraction steam from these turbines feeds the low pressure steam system. This system supplies heat for distillation columnreboilers and miscellaneous small users.

Steam Supply. As previously noted, a large portion of the plant steam is generated in the reforming section by recovering heat from both process gases and combustion flue gases. The total steam production is calculated by summing all the heat streams associated with steam generation. This calculation is performed in "DESIGN-SPEC" block STMAKE which determines total steam production from cold demineralized water at a datum temperature. (See Figure 27). The calculation is iterative, with the calculated value of steam production used to update initial guesses of boiler feed water flow rate.

Steam Demand. The calculation of the total plant steam demand uses three separate "FORTRAN" blocks - one for each steam pressure level. (See Figure 28). The calculations begin with the low pressure (LP) steam. Plant LP steam energy requirements in the form of heat and work streams are summed in block LPSTM to obtain the total LP steam demand. This material stream and energy streams associated with medium pressure (MP) steam are input to "FORTRAN" block MPSTM. The total extraction steam rate from the medium pressure steam turbines is equated to the low pressure demand. The required steam to the condensing sections of the turbines is back calculated from the known total work requirements of the turbines. Summation of steam flows to the turbines and other process units permits calculation of total demand for medium pressure steam. The total plant demand for high pressure steam is computed in an analogous procedure in "DESIGN-SPEC" block HPSTM.

Steam Supply/Demand Balancing. In most instances, an imbalance will result between calculated plant steam demand and steam generation capability. In the actual plant, this imbalance is overcome by manipulation of the auxiliary firing rate in the convection duct, or export of excess steam to other plants in our facility. In addition, the operator has some flexibility in selection of steam or electric motor drives for certain pieces of equipment. For the present, an iterative loop to balance the steam system by adjustment of these variables is not planned.

GENERAL COMMENTS ABOUT PROGRAM

ASPEN has many features which make it one of the most powerful process simulators. This section discusses our experience with these features as a Plan A[1] participant in the test program, and notes limitations and difficulties where applicable.

Since our model was developed during the ASPEN testing phase, program "bugs" were expected. Several program "bugs" were noted which prevented us from using the program in its most efficient way. However the flexibility of ASPEN permitted us to circumvent these problems. For example, in May, 1981 four alternative approaches were required for the compressor turbine section before one finally worked. Now all four approaches work.

We also found the hot-line support and ASPEN staff to be very helpful. There were many instances where the fix was a simple one or two line FORTRAN modification. For these cases the ASPEN staff would call and identify the fixes. We would update the FORTRAN program, compile it and link it into our load module. Other problems required more of a fix, and we had to wait until the next update. Still others were not corrected because of priorities and manpower limitations of the ASPEN staff. One example is the mass flowrate and density profiles which are missing from "RADFRC" output. This

---

(1) As a Plan A participant, we had the source and object ASPEN code installed on our ARCO computer.

enhancement should be relatively easy to make, but without it a column cannot be designed or rated.

The ASPEN manual left a lot to be desired during the testing phase. Parts were incomplete or even incorrect. In one instance, the same variable was given a different name in different sections. The ASPEN hot-line helped clarify many problems with the manual.

User blocks were written with relative ease. They were first written as stand alone FORTRAN programs, and then modified to be ASPEN users subroutines as dictated by the ASPEN program. Due to the required reprogramming and difficulties with the ASPEN manual, this modification took approximately two weeks for our summer employees. However, we experienced no major difficulty.

"FORTRAN" blocks are very easy to use and extremely flexible since any ASPEN variables can be addressed and adjusted. We experienced a "bug" when the total flow of more than one stream was adjusted in the same "FORTRAN" block. We circumvented the problem by using two "FORTRAN" blocks, and have not rechecked the early run.

"DESIGN-SPEC" blocks are fairly easy to use. We experienced two difficulties with the testing version. First, the limit statement only accepted constants which had to be set in the input data. The ASPEN subroutine SETLIM now permits one to adjust these limits during the calculation in another "DESIGN-SPEC" or "FORTRAN" block. Some problems were experienced with SETLIM. Second, a "DESIGN-SPEC" only resets the parameter being varied. Consequently, a referenced variable can not be reset by an F - statement in the "DESIGN-SPEC" like it can in a "FORTRAN" block. This caused some initial problems since the manual did not report this limitation.

Recycle convergence loops are very powerful especially when coupled with the printout diagnostics. These loops can be applied to "DESIGN-SPEC" or material recycle loops. They remember what happened the last time this recycle loop converged. The order of calculation default is often very helpful, but engineering judgement should override this default when necessary. Problems encountered include, not being able to use all secondary keywords, and some "DESIGN-SPEC" suddenly stop converging. The answer when the "DESIGN-SPEC" suddenly stopped converging were usually acceptable. Material recycle streams also failed to exclude trace components from their default tolerance.

Printout diagnostics are very powerful for ASPEN. One can easily increase or decrease the diagnostic level for physical properties, streams, system and simulation. This helps to identify a "bug" or minimize computer output.

Thermodynamic options available in ASPEN are extremely flexible. Built in property routes have been stored in ASPEN. The engineer can accept an entire thermodynamic option or modify parts or all of the option. The user can even input his own method as we did for the high pressure methanol-water binary.

Information streams are a major advance in process simulation. They permit the engineer to easily transfer information, like heat or power requirements, from one block to another. They also permit easy visualization of recycle loops involving heat or power.

Although ASPEN is a very powerful and useful process simulator, correcting the following deficiencies would be helpful. Stream reports are awkward in regard to component identification and the order in which streams are printed. Computer setup CPU time is substantial compared to the execution of the simulation. User blocks are presently limited to four material inlet, four material outlet streams and one information outlet stream. These should be increased, and an inlet information stream permitted.

CONCLUSION

The power and flexibility of the ASPEN simulation was demonstrated. This simulation used its full capability to develop a successful rating model of our new methanol process. New user blocks were written to describe unit operations for which no unit operation block exists. "FORTRAN" blocks and "DESIGN-SPEC" were used to develop rating models and to simulate control schemes. Proprietary thermodynamics were incorporated when existing thermodynamics were not adequate.

The paper also demonstrates the flexibility of a "layered approach simulation."

Some "bugs" were uncovered during the simulation; however, alternative approaches were developed to circumvent them.

NOTATION

A	= heat transfer area
f	= ratio of mean temperature different (MTD) to log mean temperature difference (LMTD)
M	= mass flow rate
q	= heat transfer rate
T_{CI}	= cold stream inlet temperature
T_{CO}	= cold stream outlet temperature
T_{HI}	= hot stream inlet temperature
T_{HOC}	= calculated hot stream outlet temperature
T_{HOG}	= guess hot stream outlet temperature
ΔT_{LM}	= log mean temperature difference
U	= overall heat transfer coefficient

Greek Symbols

ρ	= density

Literature Cited

Evans, L.B., J. F. Boston, H. I. Britt, P. W. Gallier, P. K. Gupta, B. Joseph, V. Mahalec, E. Ng, W. D. Seider and H. Yazi, "ASPEN : An Advanced System for Process Engineering," Computers in Chemical Engineering, 3, 319, (1979).

Stephens, A.D., "Stability and Optimization of a Methanol Converter," ICI Symposium Paper, 25, (1974).

White, III, C.W., W. D. Sieder, "Computation of Phase and Chemical Equilibrium Part IV: Approach to Chemical Equilibrium," AIChE J., 27 (3), 466, (1981).

ACKNOWLEDGEMENTS

The summer work of Greg Markiewicz and Michael Lee is appreciated. They wrote ASPEN user blocks for the compressor and reformer duct. We are also thankful for the basic compressor algorithm supplied by Gil Cruz.

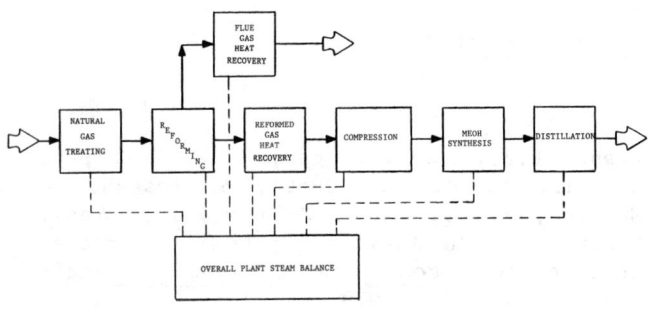

Figure 1. Simplified block flow diagram of methanol process.

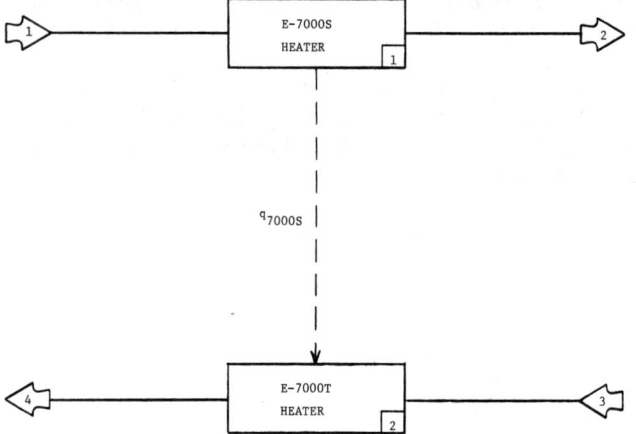

Figure 2. "Level zero" simulation of a heat exchanger.

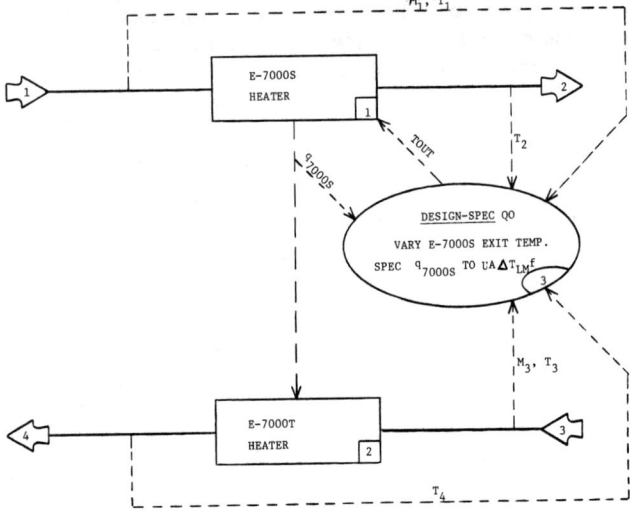

Figure 3. "Level one" simulation of a heat exchanger.

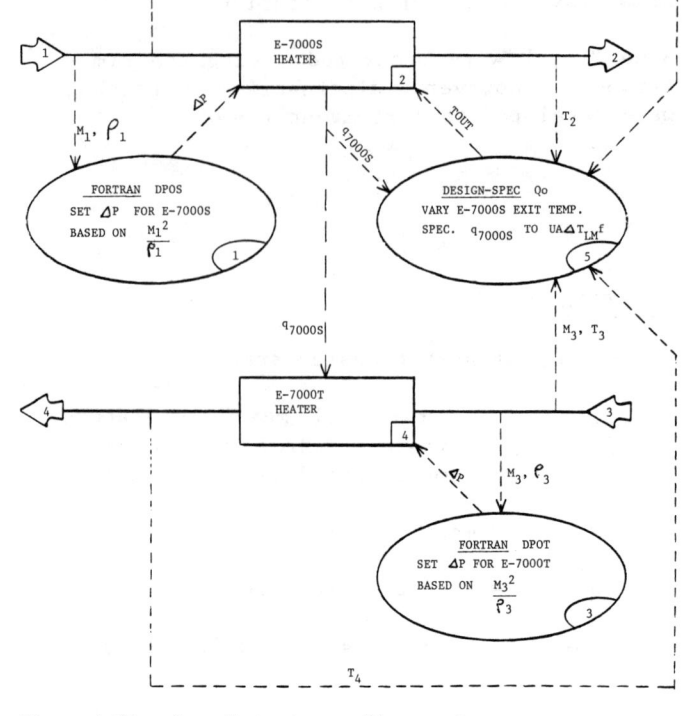

Figure 4. "Level two" simulation of heat exchanger.

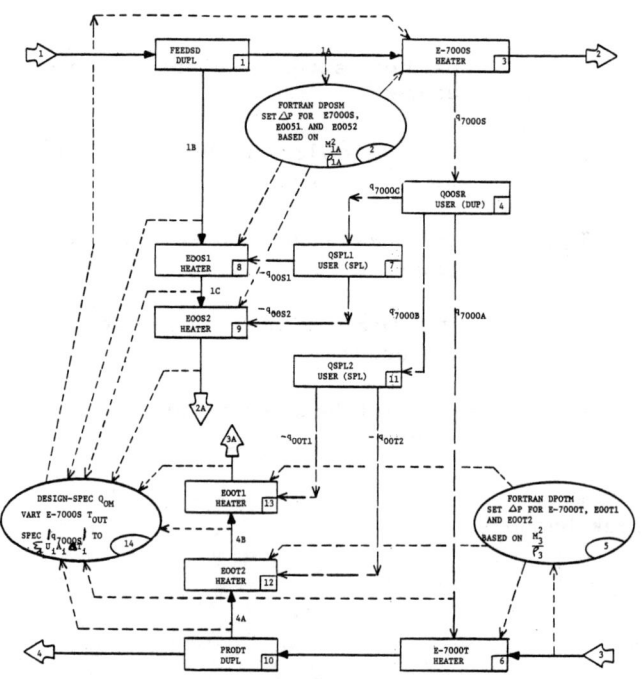

Figure 5. "Level three" simulation of heat exchanger.

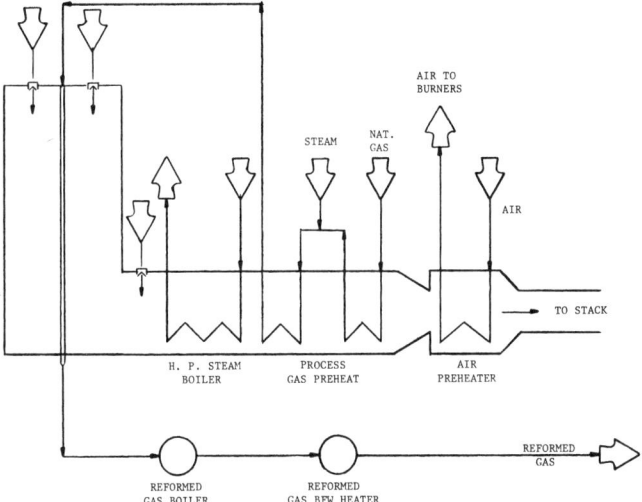

Figure 6. Schematic of methanol plant reformer.

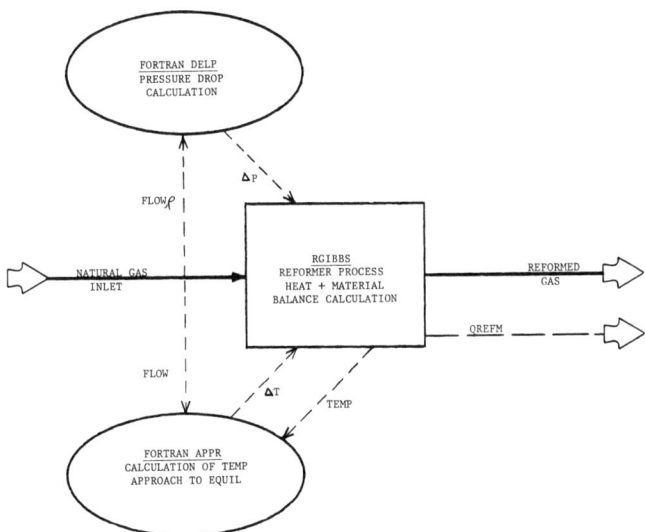

Figure 7. Reformer process side simulation.

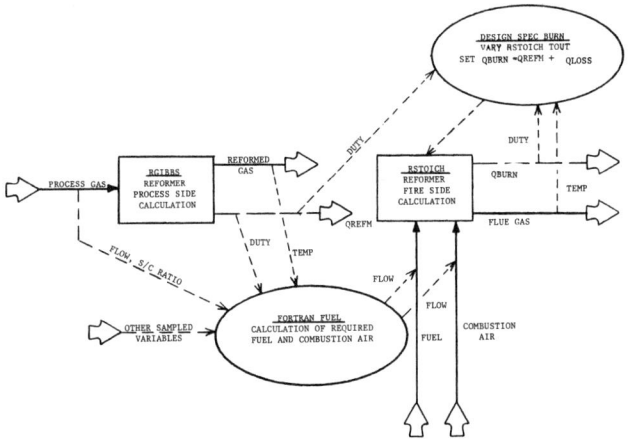

Figure 8. Simulation algorithm for reformer radiant section.

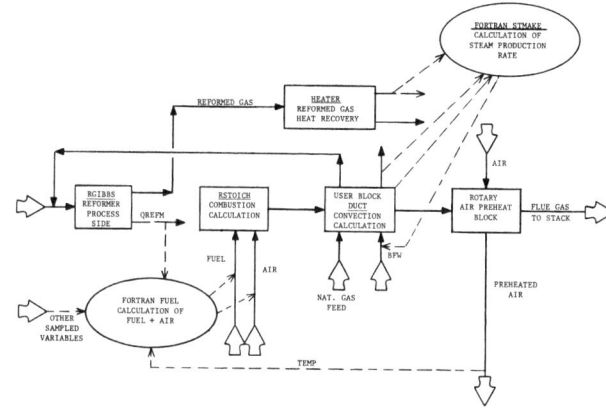

Figure 9. Simplified flowsheet of overall reformer simulation.

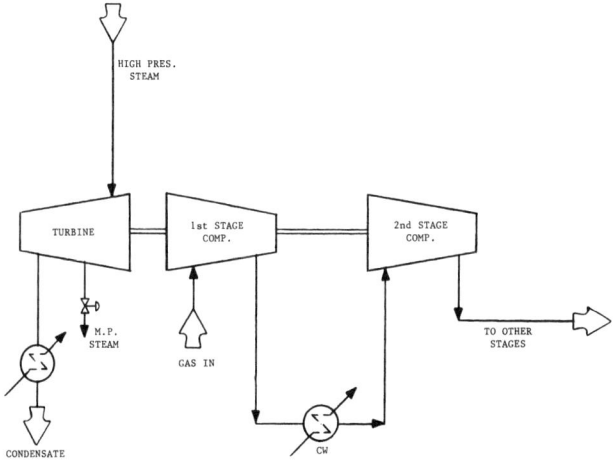

Figure 10. Schematic of compression section equipment.

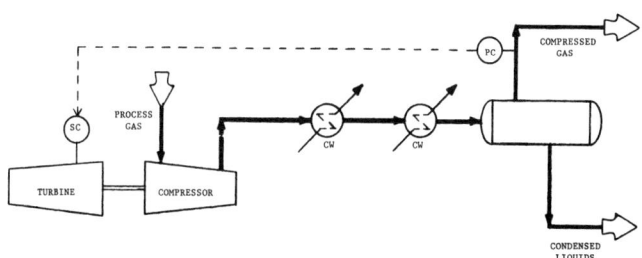

Figure 11. Typical compressor control algorithm.

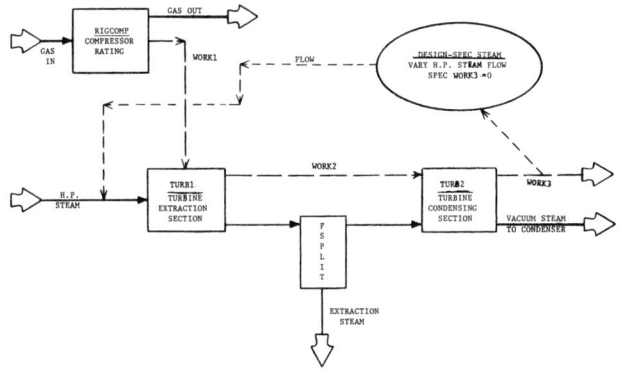

Figure 12. Simulation of compressor-turbine system.

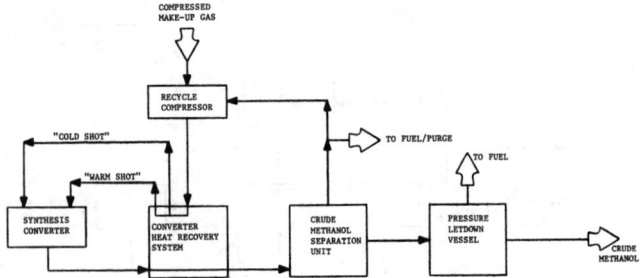

Figure 13. Simplified synthesis reactor and heat recovery loop.

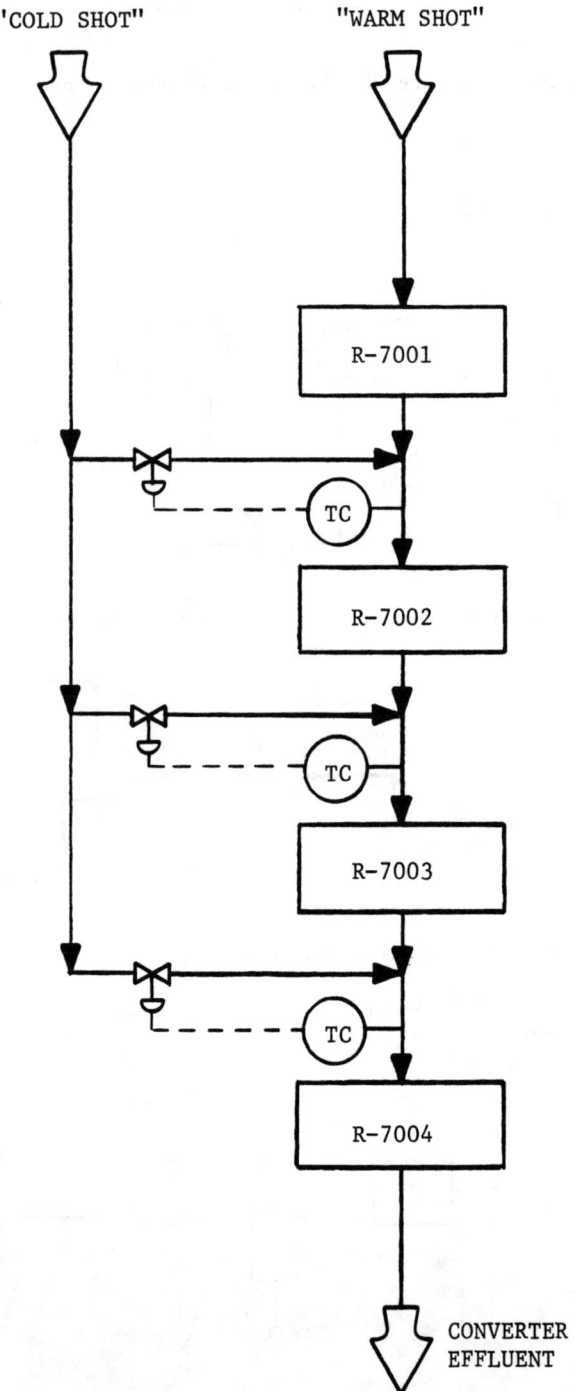

Figure 14. Schematic of four stage synthesis reactor.

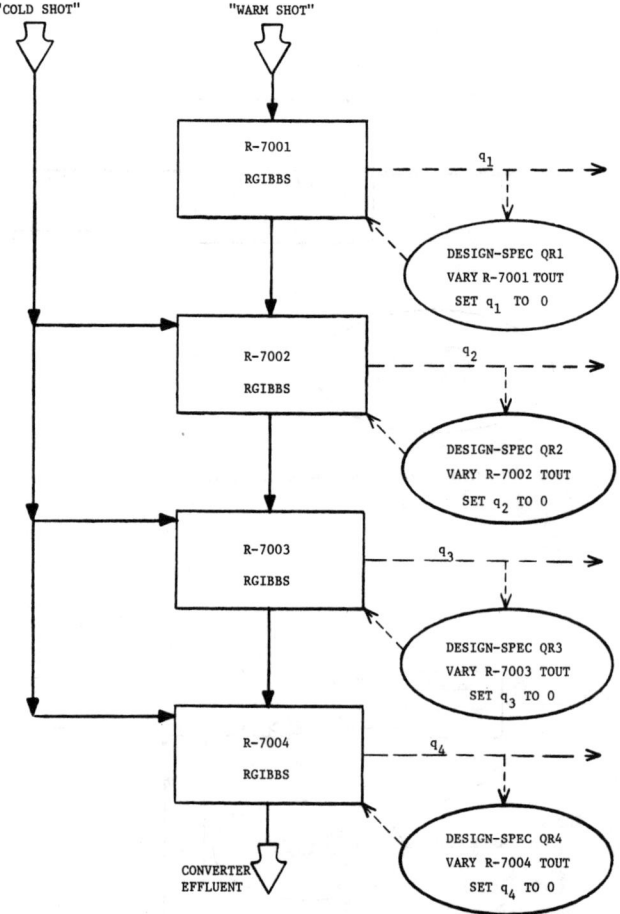

Figure 15. "Level zero" synthesis reactor model.

Figure 16. "Level one" synthesis reactor model.

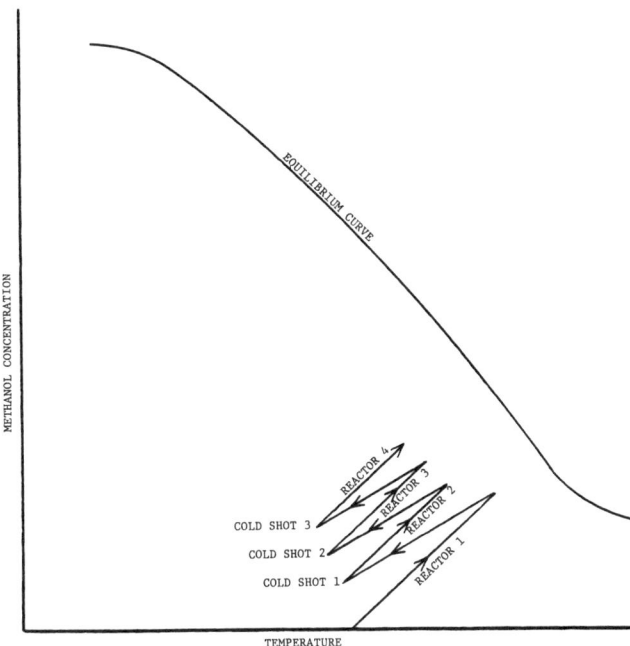

Figure 17. Reactor profile.

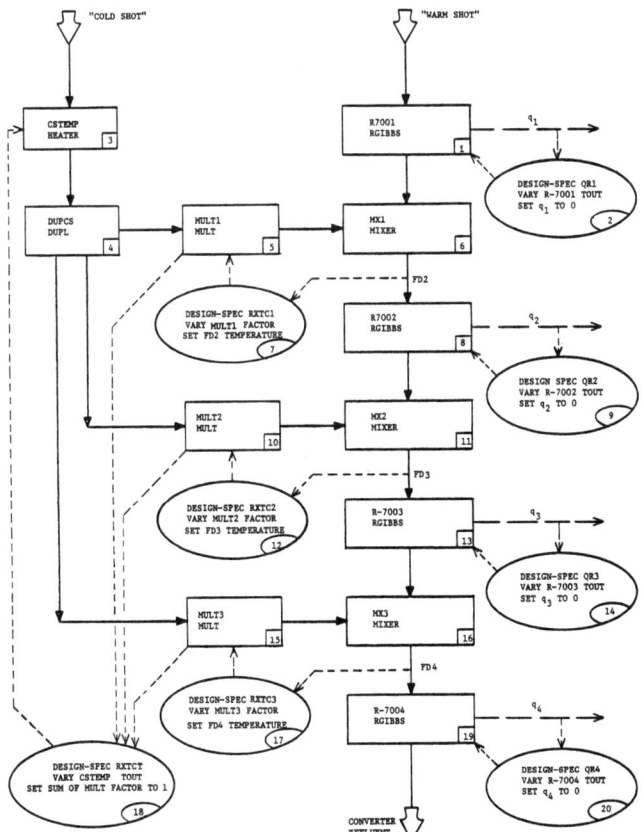

Figure 18. "Level two" methanol converter loop block diagram.

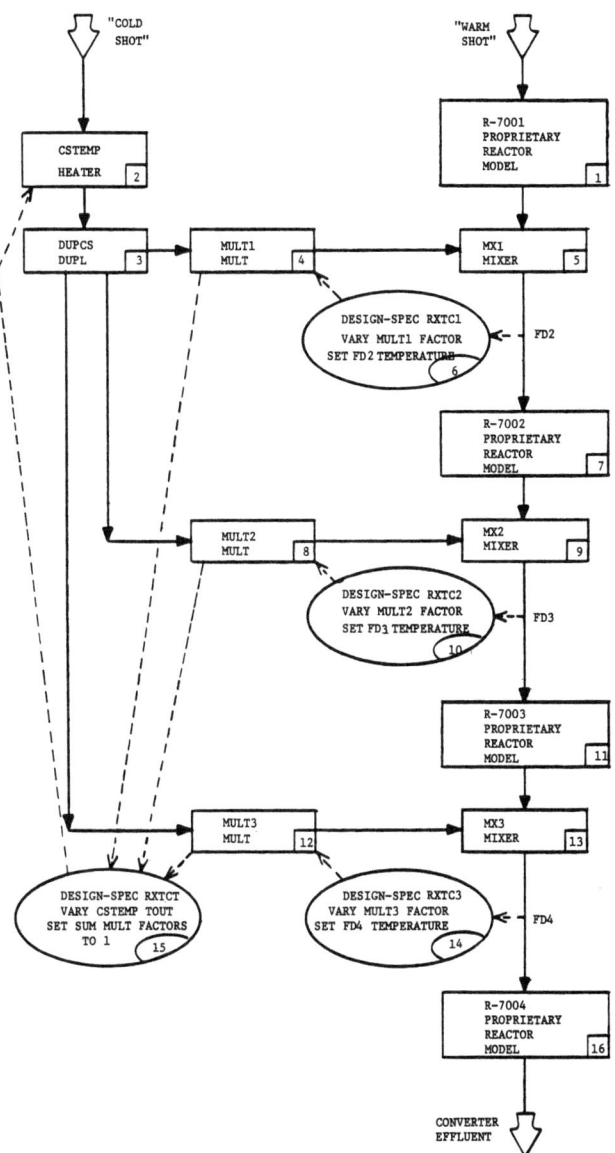

Figure 19. "Level three" methanol converter loop block diagram.

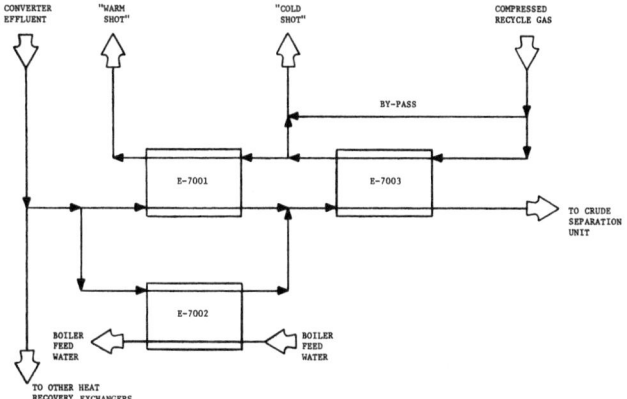

Figure 20. Converter heat recovery system.

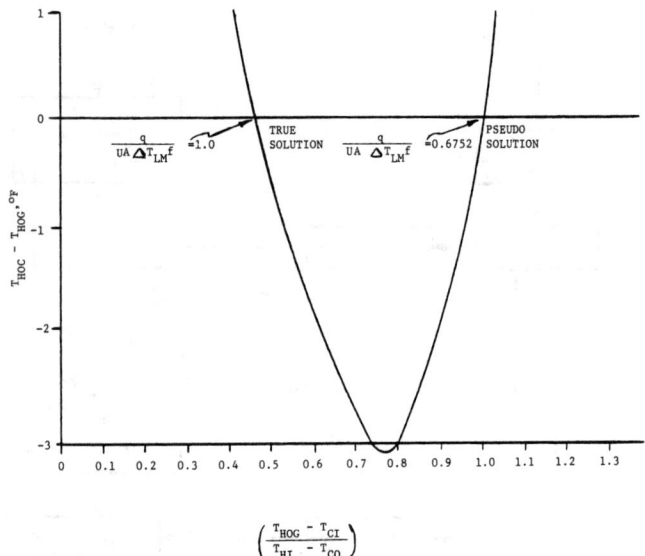

Figure 21. "Level zero" converter heat recovery system.

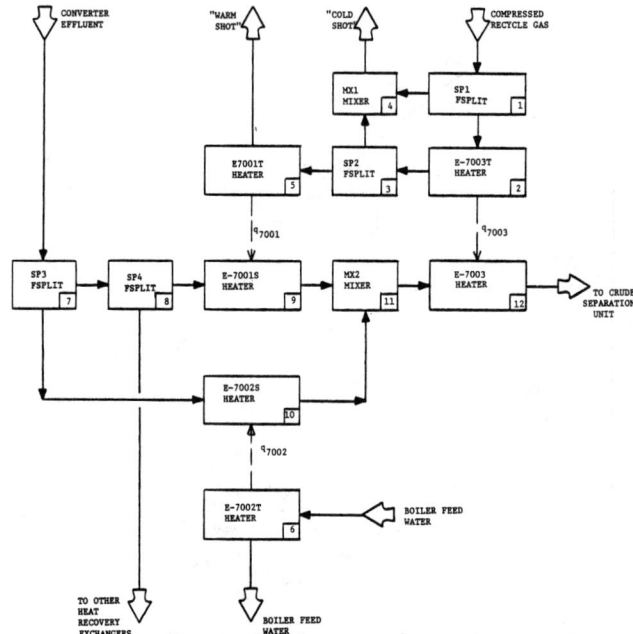

Figure 23. A solution to

$$T_{HOC} = T_{CI} + (T_{HI} - T_{CO}) \exp\left\{\frac{UAf}{q}\left[(T_{HOF} - T_{CI}) - (T_{HI} - T_{CO})\right]\right\}$$

where

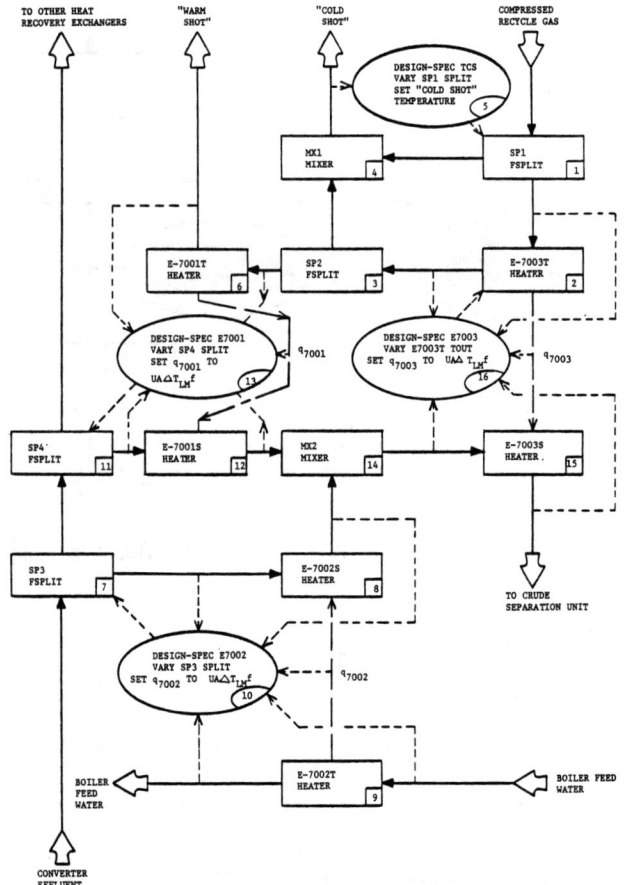

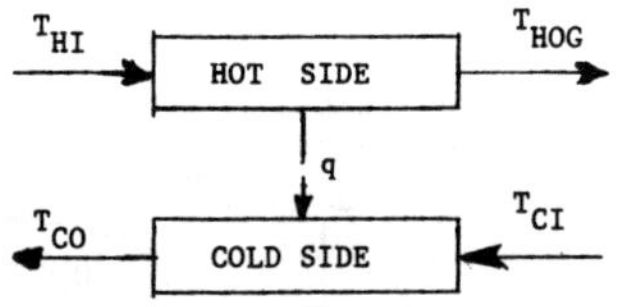

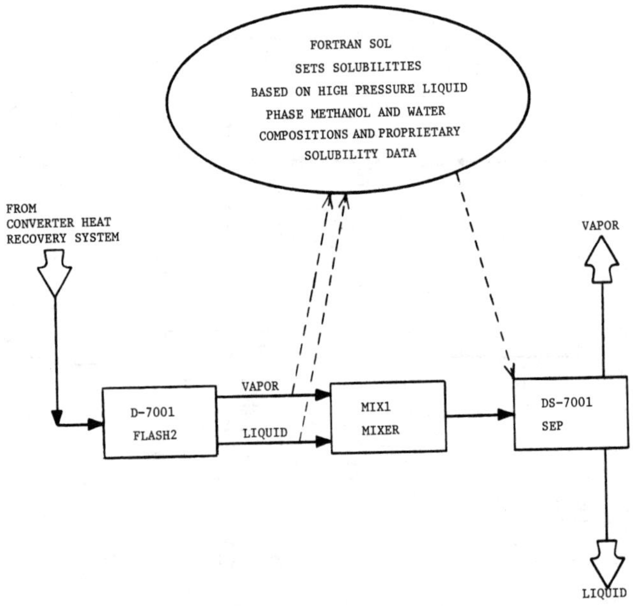

Figure 22. "Level one" converter heat recovery system.

Figure 24. Crude methanol separation unit model.

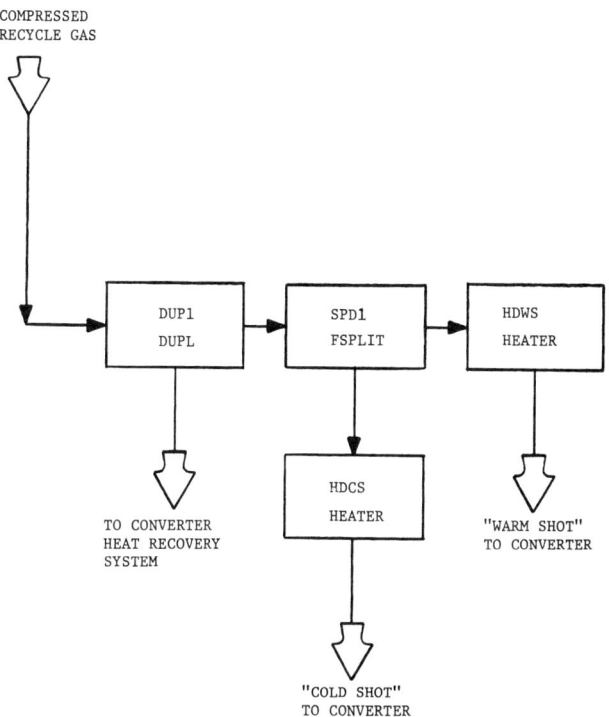

Figure 25. Alternate approach to close converter heat recovery system energy transfer loop.

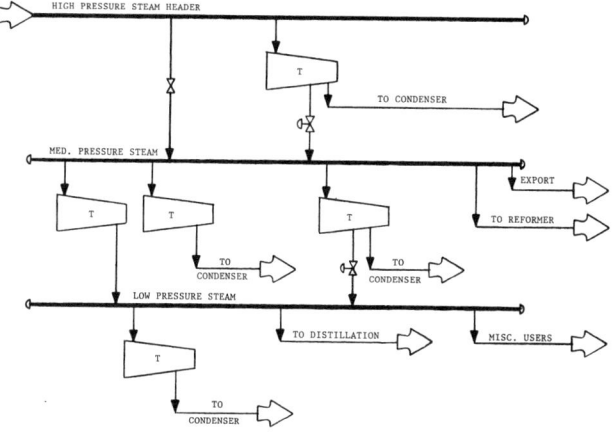

Figure 26. Simplified steam system diagram.

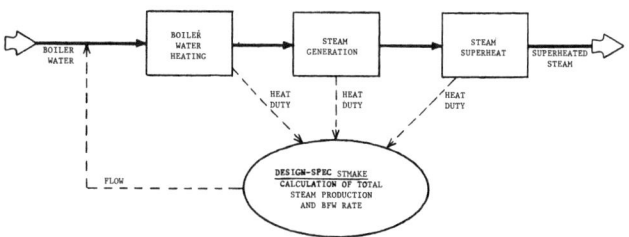

Figure 27. "Level 0" steam production algorithm.

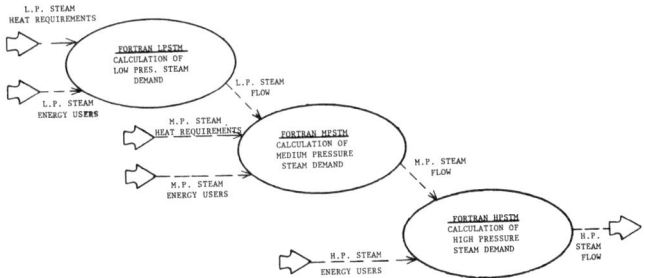

Figure 28. "Level 0" simulation of plant steam demand.

THE APPLICATION OF ASPEN FLOWSHEET SIMULATOR AT ALCOA

V.B. SHAH
P. GACKA
and
J.M. LANGA

Aluminum Company of America
Alcoa Technical Center
Alcoa Center, Pa. 15239

Bayer (Aluminum Refining) process simulation using ASPEN prompted as to develop to unit operation models which were not available in the ASPEM system. This paper described the development of a CCD (Counter Current Decantation System) model to be used in the ASPEN system as a user block, and a TURGEN (Turbine-Generator) model to be inserted as a new ASPEN block. A short description of Bayer process is also attempted. We presume that readers are familiar with the ASPEN system and its capabilities. Finally, our experience on the use of ASPEN at Alcoa is summarized.

1. INTRODUCTION

The ASPEN flow sheet simulator has been used at Alcoa for the simulation of Bayer process plants and process power plants. The advanced capabilities of ASPEN have greatly increased the efficiency of engineers developing this simulation. This paper describes the use of two ASPEN capabilities: user blocks and new ASPEN blocks. Also, a summary of other useful capabilities is given. Because of the unfamiliarity of many readers with the Bayer Refining Process, a brief description is offered.

2. BAYER PROCESS DESCRIPTION

Bayer process produces purified aluminum oxide (alumina) from bauxite ores. The fundamental steps of the process are: extraction of the bauxite in a sodium hydroxide solution (digestion), separation of the undissolved bauxite residue in clarification and the precipitation of the pure alumina trihydrate crystals, these crystals then calcined to give the product in its final form.

Figure 1 shows a schematic of the process. In digestion, ground bauxite is mixed with recycled (spent) caustic solution. This slurry is heated and the alumina is dissolved. The dissolution reaction equation is:

$$Al_2O_3 - 3H_2O + 2NaOH \rightleftarrows 2 NaAlO_2 + 4H_2O$$

This reaction takes place at 290°F and approximately 60 psia. Next, the insoluble bauxite residue is separated from the rich liquor. This separation is done in a series of mixing, settling and washing operations. The combination of these operations is called clarification. After cooling, the liquor is then seeded to start the precipitation.

The crystals are classified by size with the smaller particles recycled as seed and the larger particles sent to calcination. Regenerative heat exchange occurs between hot and cool liquor streams in digestion and heat interchange to recover energy. Evaporation is required to counteract the dilution of the liquor, which occurs largely in the clarification department.

The process is energy intensive. Where possible, power plants have been built in connection with Bayer plants to cogenerate power and steam for the process. Simulations of both the process plant and the powerhouse were developed to predict the effect of changes in plant design or operating conditions upon the operation of the powerhouse.

3. APPLICATION OF USER BLOCKS, THE COUNTER-CURRENT DECANTATION MODEL

One of the major unit operations in the clarification department is the counter-current decantation (CCD) system. The operation separates residue from the liquor without losing large amounts of caustic and alumina. The residue is disposed in the mud lake. CCD is considered a multistage mixer and settler system (see Figure 2), with feed streams of slurry (solid residue and highly concentrated liquor) and lake water (liquor of very low concentration). These streams are mixed and allowed to settle. The outputs of the settling operation are an overflow stream of decanted clear liquor and an underflow slurry with a low concentration of alumina and caustic. Common additions to this operation are side and withdrawal streams. A side stream is an extra input stream to a stage in the CCD. A withdrawal stream occurs when a fraction of an interstage overflow or underflow is removed for use elsewhere in the process. These streams will be considered inputs and outputs of the system, respectively.

The CCD system could be modeled by two different approaches. One would involve the use of available ASPEN unit operation models and Fortran blocks. The other option involves the development and use of a user block. The ASPEN block diagram for a model of CCD, using standard unit operation models and Fortran blocks, is shown in Figure 3. Each stage is made up of a mixer block, a component splitter block and a Fortran block.

The mixer would be used to combine all side streams, wash streams and mud feed streams. The Fortran block does the calculations needed to determine the underflow and overflow split for the splitter block. Details of the calculations will be covered later. Modeling CCD this way allows the user complete flexibility with regard to number of side streams, for any particular stage. A user would have to add another material stream as input to the mixer. If the user desired to withdraw part of an overflow of underflow from the CCD, a flow splitter (FSPLIT) block could be added in the appropriate place.

However, there are a number of reasons for developing a user block. Most Bayer process engineers like to think of the CCD system as being one unit operation and would be more satisfied with a model which described the system as one unit. A user block is a convenient way of modeling CCD as one unit operation. If the ASPEN unit operation model approach was used, these users would not be able to change the number of washing stages or move a sidestream and/or withdraw from one stage to another without adding new blocks, Fortran blocks, changing stream names, and adding tear streams. This has the possibility of forcing a change in the convergence sequence. Such changes would be too involved for users with little or no ASPEN knowledge. However, if CCD was modeled as a user block, such changes could be made by varying one parameter in the user block input language. Also, as a user block, CCD could be modeled more efficiently. The mass balance could be solved directly, as compared to the iterative convergence scheme used by the first method.

It was decided to develop a user block for CCD to exploit these advantages, but it was developed so that flexibility could be maintained in regard to sidestream and withdraw stream placement. Next, the algorithm used for the CCD user block (USRCCD) will be described.

A. CCD Algorithm

Figure 4 shows a typical internal stage of CCD system. Terms used in the diagram are:

$U_n, \underline{x}_n$ Total flow and composition of underflow from stage n.

$U_n, \underline{x}'_n$ — Total flow and composition of underflow from stage n after adjusting for side feed and withdrawal.

$O_n, \underline{X}_n$ — Total flow and composition of overflow from stage n.

$O_n', \underline{X}'_n$ — Total flow and composition of overflow from stage n after adjusting for side feed and withdrawal.

W_n^u — Total withdrawal rate from underflow at stage n.

W_n^o — Total withdrawal rate from overflow at stage n.

$F_n^s, \underline{f}_n^s$ — Total flow and composition of feed at stage n which is completely mixed with the underflow from stage above.

$F_n^c, \underline{f}_n^c$ — Total flow and composition of feed at stage n which is completely mixed with the overflow from stage below.

Equations describing the operation of a typical internal stage n are described below. The component index notation is suppressed for simplicity. The total number of equations of each type is indicated in parentheses:

B. <u>Total Mass Balance (2N)</u>

$$U_n = TS/PS \quad (1)$$

$$-(U_{n-1} - W_{n-1}^u) - (O_{n+1} - W_{n+1}^o) + O_n + U_n = F_n^c + F_n^s \quad (2)$$

where: TS is total solid flows through the system PS is solid fraction of underflow.

C. <u>Component Mass Balance (CN)</u>

$$-(U_{n-1} - W_{n-1}^u)x_{n-1} + x_n U_n + X_n O_n - (O_{n+1} - W_{n+1}^o)X_{n+1} = f_n^s F_n^s + f_n^s F_n^s \quad (3)$$

D. <u>Imperfect Mixing (CN)</u>

Scandrett (1963) defined mixing efficiency at any stage as:

$$E_n = (x'_{n-1} - x_n)/(x'_{n-1} - X_n) \quad (4a)$$

where: $x'_{n-1} = (U_{n-1} - W_{n-1}^u)x_{n-1}/(U_{n-1} - W_{n-1}^u + F_n^s) + f_n^s F_n^s / (U_{n-1} - W_{n-1}^u + F_n^s)$.

By definition, E_n is a function of stage number only and should be the same for all components. This allows equation 4a to be written as,

$$X_n = \frac{1}{E_n} x_n - \left(\frac{1}{E_n} - 1\right) \left[\frac{(u_{n-1} - W_{n-1}^u)x_{n-1} + f_n^s F_n^s}{u_{n-1} - W_{n-1}^u + F_n^s}\right] \quad (4)$$

E. <u>Energy Balance (N)</u>

$$-(U_{n-1} - W_{n-1}^u)H_{n-1}^u + U_n H_n^u + O_n H_n^o - (O_{n+1} - W_{n+1}^o)H_n^o = H_n^c F_n^c + H_n^s F_n^s \quad (5)$$

where:
H_n^u — Enthalpy of underflow from stage n.
H_n^o — Enthalpy of overflow from stage n.
H_n^c — Enthalpy of clear feed.
H_n^s — Enthalpy of solid feed.

The algorithm for solving these equations is simplified by the assumption that the mixing efficiency on each stage is fixed and independent of temperature and composition. Also, the percent solids in the underflow at each stage is fixed by the user. These two assumptions decouple the solution of the equations 1 through 5. This allows us to solve the equations as shown in Figure 5. The total mass balance equations, being independent of either compositions or temperature, are solved first.

The next step is to solve component mass balance and imperfect mixing equations simultaneously. Substituting equation 4 in 3 yields, a set of tridiagonal equations.

$$B_1 x_1 - C_1 x_2 = F_1$$

$$-x_{n-1} + B_n x_n - C_n x_{n+1} = F_n \quad (6)$$

$$-x_{N-1} + B_N x_N = F_N$$

where:

$$B_1 = (U_1 + O_1 R_1 - (R_2-1)(O_2 - W_2^o)$$

$$(U_1 - W_1^u)/y_1)/D_1$$

$$C_1 = (O_2 - W_2^o) R_2 / D_1$$

$$F_1 = (f_1^c F_1^c + f_1^s F_1^s - (R_1-1) O_1 f_1^s +$$

$$(R_2-1)(O_2 - W_2^o) f_2^s F_2^s / y_1)/D_1$$

$$D_1 = F_1 + O_1(R_1-1)$$

$$B_n = (U_n + O_n R_n - (R_{n+1}-1)$$

$$(O_{n+1} - W_{n+1}^o)(U_n - W_n^u)/y_n)/D_n$$

$$y_n = U_n - W_n^u + F_{n+1}^s$$

$$C_n = (O_{n+1} - W_{n+1}^o) R_{n+1} / D_n$$

$$F_n = (f_n^c F_n^c + f_n^s F_n^s - (R_n-1)$$

$$O_n f_n^s F_n^s / y_{n-1} + (R_{n+1}-1)(O_{n+1} - W_{n+1}^o)$$

$$\cdot f_{n+1}^s F_{n+1}^s / y_n)/D_n$$

$$D_n = U_{n-1} - W_{n-1}^u - (R_n-1) O_n$$

$$R_n = 1/E_n$$

Equation 6 is in convenient tridiagonal form and can be solved using Thomas algorithm as documented by Grabble et al (1958).

Once the total mass balance and component mass balances have been solved, the next step is to solve the energy balance equations.

The assumption is made that underflow and overflow temperatures for a particular stage are the same. This temperature is referred to as the "stage temperature." The calculation starts from the "top" down (see Figure 6). The first stage temperature is estimated, and from that estimate underflow and overflow enthalpies are calculated. An energy balance is performed on the first stage and the overflow enthalpy of the second stage is calculated. The second stage overflow temperature (the second stage temperature) is calculated from that enthalpy. An energy balance is performed on the second stage to calculate third stage overflow enthalpy and temperature. This sequence is repeated until the last stage is reached. The "next stage overflow" in this case is the input wash water stream. The calculated temperature is compared with the input temperature of the wash water stream. If the difference between these two is larger than a prescribed tolerance, the first stage temperature is adjusted using a one dimentional equation solving routine and the calculation sequence continues until convergence is met. Special considerations are made for the case of complete withdrawal of the next stage overflow, so that no enthalpy calculations are done for a stream of zero flow.

This algorithm has been implemented as a ASPEN user model (USRCCD). USRCCD has been tested extensively with the Bayer process simulation.

4. APPLICATION OF NEW BLOCK TO POWERHOUSE SIMULATION

As described earlier, a powerhouse is an integral part of the Bayer refining process. Figure 7 shows a simplified schematic of a typical steam and power generation process. The process involves boiling and superheating the feed water at high pressure. This high pressure steam passes through turbines, producing shaft power which is converted to electrical power and ejecting steam at low pressures to be used in the refining process.

The ASPEN block diagram for this process is presented in Figure 8, without detailing the turbine generator units. The ASPEN block diagram for a typical two-stage turbine with generator is outlined in Figure 9 using standard ASPEN blocks which do not include an electrical power generator block.

Providing a customized generator block does not save much effort on the part of a process engineer in defining ASPEN input (Figure 10). This leads to the development of a new ASPEN block called TURGEN.

As designed, TURGEN can handle three-stage turbines with extraction at any stage. Use of this new block will reduce the ASPEN flow diagram to a single block, as shown in Figure 11. The function of this block and its method of solution are described next.

A. <u>TURGEN</u>: Multistage turbine generator system.

1. Function

TURGEN simulates the operation of a multistage turbine driving a generator shaft to produce electrical power. This block has three modes of calculation:

(a) The calculation of shaft power for each turbine given one of the following: isentropic efficiency, outlet temperature or outlet enthalpy. From the shaft power and generator losses (or efficiency) this routine calculates generator power output.

(b) The calculation of generator power output from the manufacturer's curves for power generation versus inlet and extraction flows. The calculation of shaft power from turbine data is then followed by computation of generator efficiency.

(c) The calculation of generator power output from the manufacturer's curves for power generation versus inlet and extraction flows. Turbine calculations are not performed.

This block requires one material inlet stream and from one to three material outlet streams. The model can have a power outlet stream.

2. Calculation Method:

For mode (a) the shaft power generated by each turbine is computed using the algorithm in Figure 12. The sum of these shaft powers gives the power input to the generator. Power output from the generator is determined using equation 7 or equation 8,

$$PWO = (PWI - FL)/(VL + 1) \qquad (7)$$

$$PWO = PWI*EG \qquad (8)$$

where:
- PWO = power output from generator
- PWI = shaft power to generator
- FL = fixed losses
- VL = variable losses as a fraction of output power
- EG = generator efficiency

For mode (b) the shaft power to the generator (PWI) is computed as described for mode (a). The electrical power output (PWO) is found from manufacturer's tables, with generator efficiency determined from equation 2. Efficiency should be in the range of 0 to 1, otherwise a warning is issued.

For mode (c) power output from generator, using manufacturer's tables, is computed and outlet stream conditions are set to user specified values.

B. Use of Tables

Manufacturers of turbine generator systems usually provide data for the complete system rather than turbine and generator units separately. This means that the turbine generator block should be able to read in the table of power generation versus feed and extraction flows at each stage. An interpolation routine is used to compute the value of any variable in the table when values for the remaining variables are provided.

Since there are several identical units of turbine-generators in the powerhouse, which use the same table, it would be ideal to input the tables outside of the TURGEN block and then refer to this table in the block. No plans are available at this time on how to implement this feature.

1) Compute inlet stream temperature (enthalpy (HI) known).

2) Compute inlet stream enthopy.

3) Isentropic entropy = inlet entropy.

4) From outlet pressure and isentropic entropy, compute isentropic temperature and isentropic entropy (HS).

5) Compute actual outlet enthalpy (HO) as follows:

    KODE = 1, HO provided by user
    KODE = 2, Outlet temperature provided compute enthalpy.
    KODE = 3, Isentropic efficiency (ES) provided

    $$HO = HS*ES + HI(1-ES)$$

6) From outlet enthalpy, calculate outlet temperature and entropy.

7) Shaft power = (HI - HO)*(mechanical efficiency)

Figure 12: Stepwise algorithm for calculating turbine shaft power.

5. OUR EXPERIENCE WITH ASPEN

Alcoa's main flowsheet oriented process is the alumina refining process (Bayer Process). In this process, solids are almost always present with fluid to affect the energy balance solutions. This eliminated most of the public version simulators from our consideration. ASPEN with solid handling capabilities, along with the capability to characterize nonconventional components like bauxite, was the logical choice. Our physical property correlations were not set up to be handled by a flowsheet simulator. For example, vapor pressure data were available as the boiling point rose, which is a function of caustic concentration expressed as gms/liter. We used ASPEN's DRS system to develop Wilson coefficients for this purpose. For enthalpy calculations we were able to use our own subroutines to calculate departures from water properties.

Another hurdle in our effort to model the Bayer process was the use of controls throughout the process. Even though other simulations have design specification capabilities, they restrict a control variable to be one of the process or block variables. ASPEN, with Fortran capabilities in the design specifications, can handle combinations of process or block variables as control specifications.

The Bayer process requires modeling of some unit operations (e.g. CCD) which were not provided by M.I.T. Being a table driven system, ASPEN requires adding a table entry to the system for every unit operation model to be added. It also requires writing a model interface to connect the equation solving model routine to the ASPEN executive routine. This not only demands excessive work, but also requires very good ASPEN knowledge. ASPEN compensated for this by providing a standard interface and table entry for the subroutines to be added by the user. We made extensive use of this capability, and now have nine user models including the one described earlier. In the Bayer process, recycle flows are about ten times the raw material flow which tends to slow down the convergence of the recycle loops. Although nesting was inevitable in some places, we were able to eliminate nested loops in many places by using the feedforward Fortran capabilities. ASPEN's state of the art convergence routines added stability and computational efficiency to the recycle loops.

The main problem with ASPEN was the difficulty in learning ASPEN in order to make maximum use of its capabilities. This confined our manpower to the persons who were trained in ASPEN.

6. CONCLUSIONS

Along with the capabilities of writing user blocks and new blocks, there are other features of ASPEN which are important in our simulation development. Physical property user models and user defined physical property option sets are useful, since the ASPEN data bank does not have any information on the components found in Bayer liquors. Also, the solid components in the process were represented by non-conventional components. The simulation of the Bayer process would have been impossible without these capabilities.

The ability to write in-line Fortran in design specifications and Fortran blocks was used extensively. ASPEN's state-of-the-art convergence methods performed well in terms of stability as well as computational efficiency. This is very important in the simulation of a large process.

7. REFERENCES

Grable, E.M., Ramos and Woodridge, D.E., "Handbook of Automation, Computation and Control." Vol. 1. John Wiley and Sons, Inc., New York (1958).

Scandrett, H.F., "Equations for Calculating Recovery of Soluble Values in a Countercurrent Decantation Washing System", Extractive Metallurgy of Aluminum, Volume 1, New York - London - Sidney, Interscience, (1963).

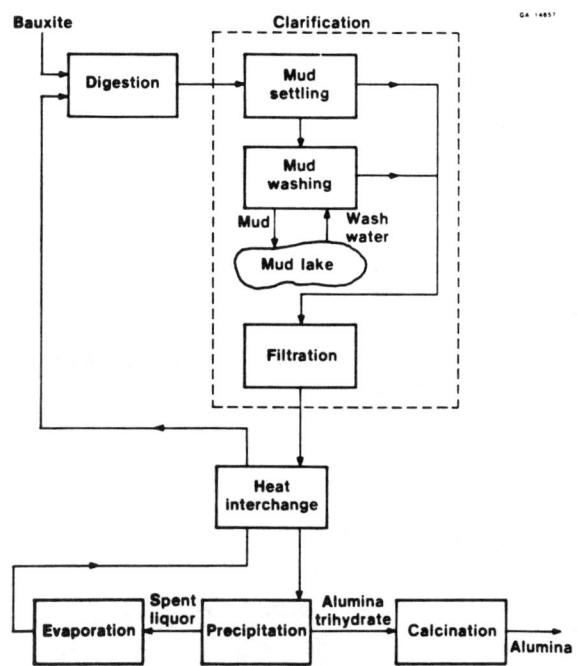

The Bayer Process Flowsheet
Figure 1

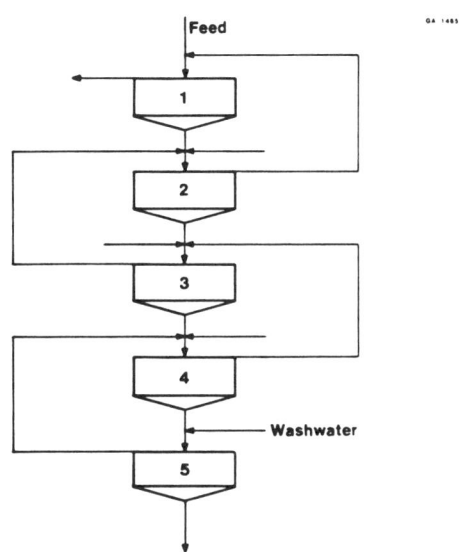

A Schematic of a 5-Stage CCD Showing Sidestreams, Feed and Washwater Streams
Figure 2

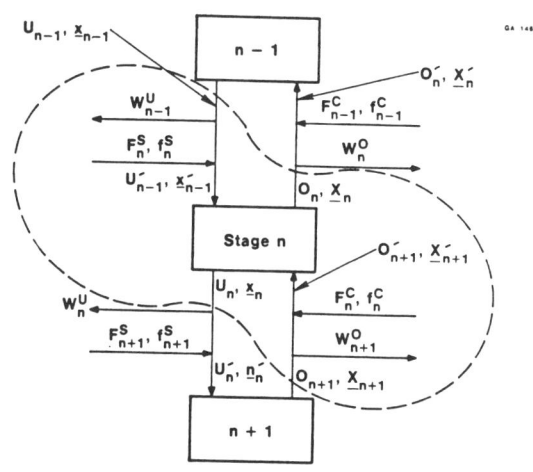

A Typical Internal Stage of CCD System
Figure 4

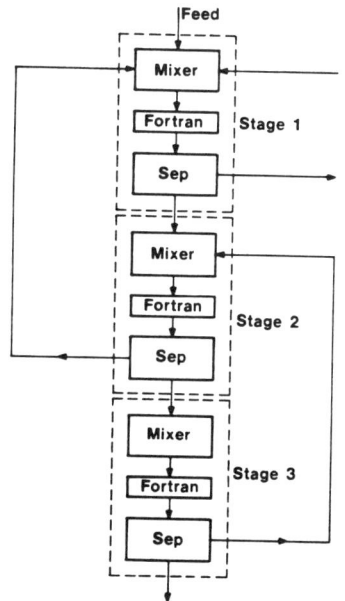

An Example of a 3 Stage Model of CCD, Modelled Using Aspen Unit Operation Blocks
Figure 3

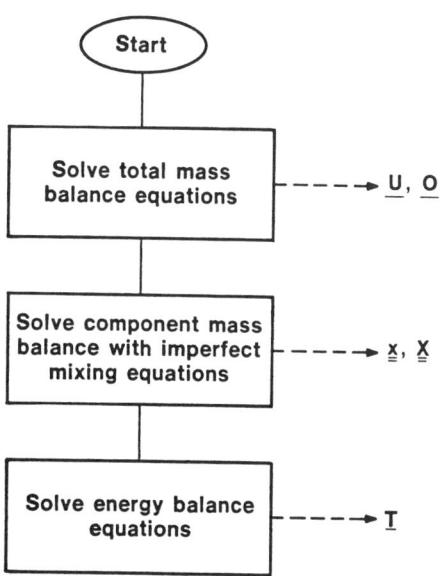

Decoupling Solution of CCD Equations by Equation Type
Figure 5

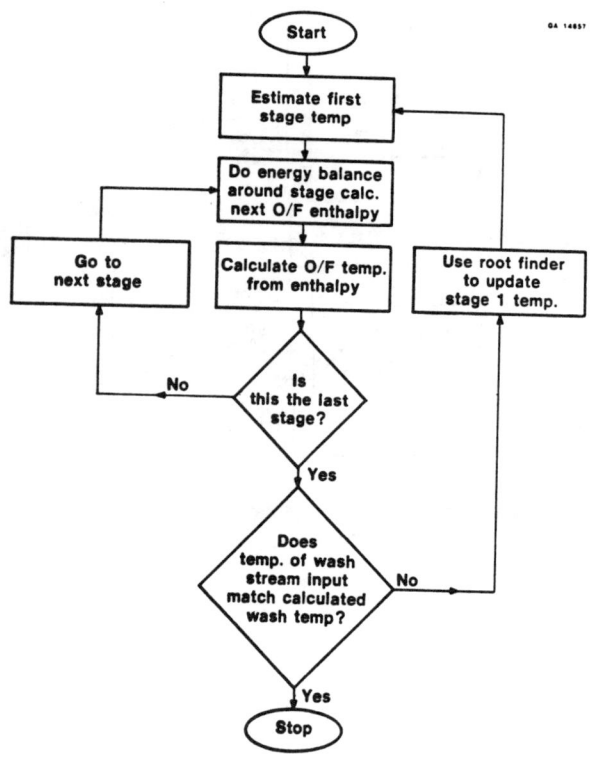

Flowsheet for Heat Balance Algorithm Used in USRCCD
Figure 6

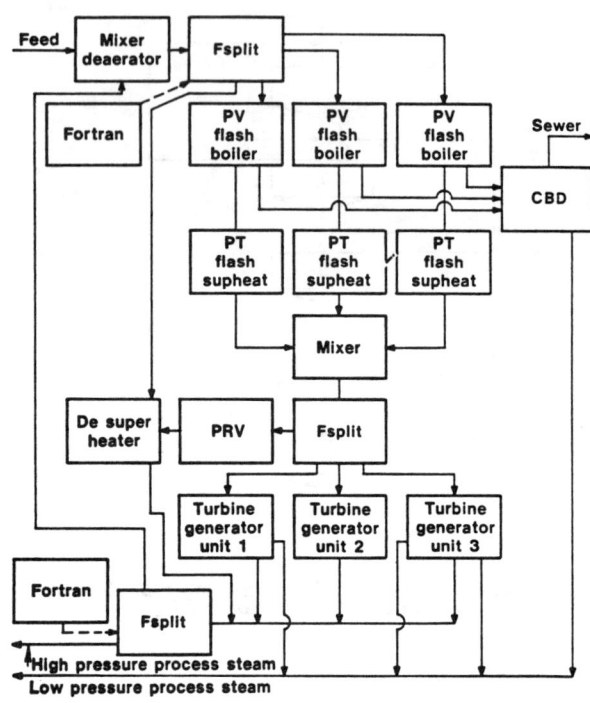

Aspen Flow Diagram of Powerhouse Process
Figure 8

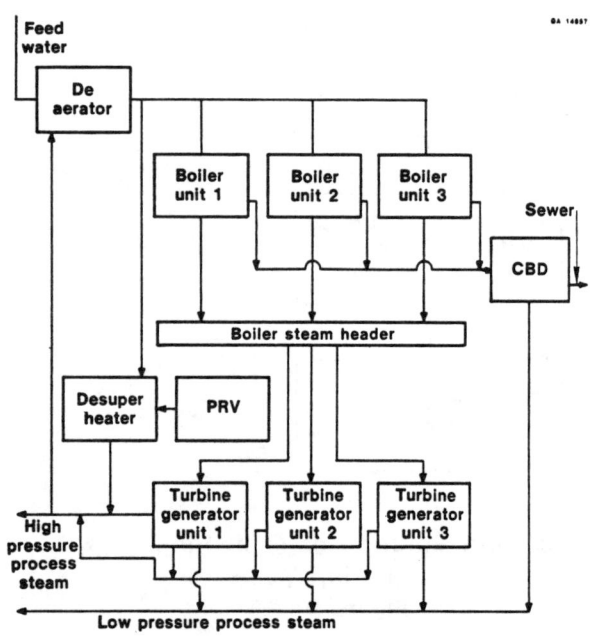

Simplified Powerhouse Schematic
Figure 7

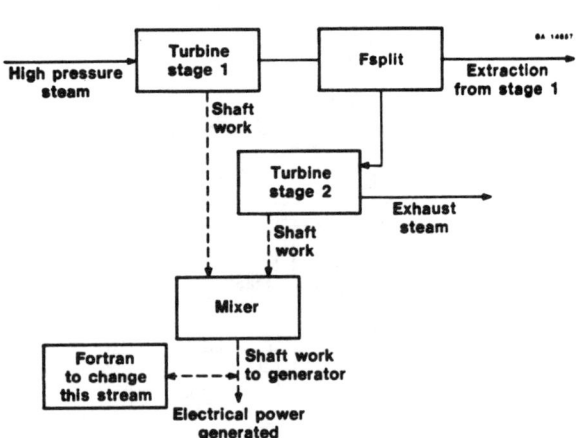

Two Stage Turbine Generator Unit Using Standard Aspen Blocks
Figure 9

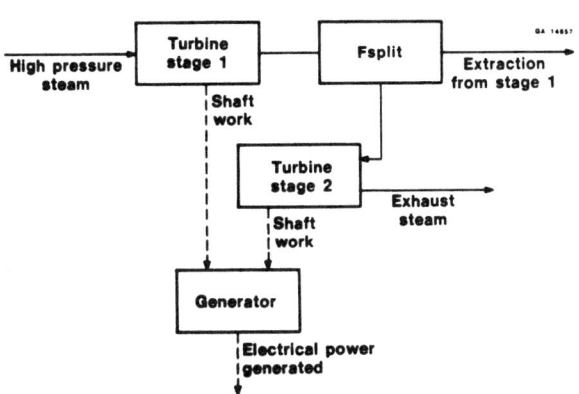

Two Stage Turbine Generator Unit Using Standard Aspen Blocks and a Generator Block
Figure 10

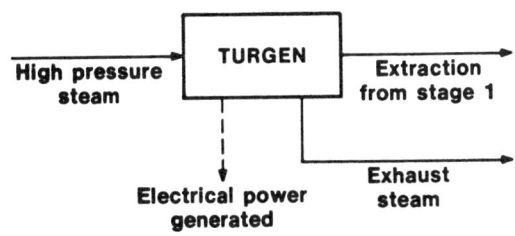

Two Stage Turbine Generator Unit Using Proposed TURGEN Block
Figure 11

THE USE OF ASPEN IN THE ANALYSIS OF THERMODYNAMIC CYCLES

L.W. FISH
D. R. EVANS
and
W.W. MADSEN

EG&G Idaho, Inc.
Idaho Falls, ID

The ASPEN computer program has been used to model a system that generates electrical power using a low-temperature binary Rankine cycle.

The equipment modeled by ASPEN consisted of a turbine, a pump, a condenser, a working fluid preheater, and a boiler. The thermodynamic properties of the working fluid were computed using the conformal solution theory version of the Benedict-Webb-Rubin equation of state.

Four thermodynamic cycles were considered, with isobutane, propane R-12, and R-22, respectively, as working fluids. The net cycle power and thermal efficiencies were essentially equivalent for the four cycles, with isobutane giving the best overall thermodynamic results. However, a preliminary economic evaluation, based on investment per kilowatt, revealed that a system using R-12 as a working fluid would be significantly less expensive than one based on the other fluids.

It was concluded that ASPEN is a useful tool for thermodynamically and economically evaluating Rankine cycles for power production. Furthermore, thermodynamic performance alone, whether based on an elementary first law analysis or on a more sophisticated second law analysis, does not necessarily define an economically optimum system design.

Until the advent in 1973 of high energy costs in the United States, there was little incentive to produce electricity or other high grade forms of energy from low-temperature (375-500 K) heat sources. In recent years, the rapidly-escalating price of energy in the United States relative to the fixed costs of investment has rendered the economics of power generation from low temperature sources considerably more favorable. Because the choice of a Rankine cycle working fluid can affect the usable power production by a factor of two or more, there is a need for a flexible analytical tool that can be used to screen working fluids and optimize the operating conditions. Of considerable advantage would be the capability to screen and optimize on the basis of the economic feasibility of the various investment options considered.

ASPEN would appear to be such a tool. Accordingly, when a recent project was funded at the Idaho National Engineering Laboratory (INEL) to design and prepare a cost estimate for a Rankine cycle to recover energy that is presently discarded during operation of the Advanced Test Reactor (ATR), ASPEN was used to perform a preliminary analysis and cost estimate. This paper will discuss this study and offer conclusions as to the usefulness of ASPEN in performing such studies.

DESCRIPTION OF EQUIPMENT

Figure 1 presents a flow diagram of the proposed ATR energy recovery system. As can be seen, the system consists of a pump, a turbine, and four heat exchangers. The need for the superheater and the regenerator is generally based on whether they increase cycle efficiency enough to offset the fixed costs of the added heat exchangers. Thus, the system configuration itself can be subject to economic studies.

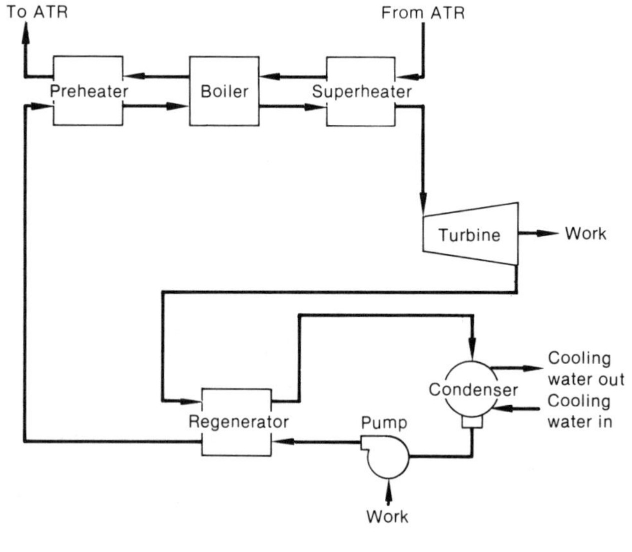

Figure 1. ATR energy recovery system

The heat source for the system is sensible heat from the primary coolant loop of the ATR. Constraints on the operation of ATR determined the entering temperature of the heat source and the total energy removed.

0065-8812-82-6400-0214-$2.00.
© The American Institute of Chemical Engineers, 1982

Also specified was the temperature of the cooling water entering the condenser.

ASPEN MODELING CONSIDERATIONS

The equipment constituting the proposed power recovery facility was modeled using the ASPEN unit operation blocks COMPR, PUMP, and HEATER. As could be surmised, the COMPR block models a compressor, which was operated in the turbine mode. The pump was assumed to be a centrifugal pump and was modeled using the PUMP block. The HEATER block was used to model the condenser, the regenerator, the preheater, the boiler, and the superheater. For sensitivity studies involving removal of the regenerator or superheater, the heat transfer rate and working fluid pressure drop for the heat exchanger in question would be set to zero.

Flowsheet convergence is needed for this system because of the feedback of heat in the regenerator and because of the working fluid loop being closed. With temperatures for the source and the sink, working fluid pressure drops in the heat exchangers, and working fluid mass flow rate specified, the working fluid stream was torn at the inlet of the turbine to allow convergence.

Two hydrocarbons (propane and isobutane) and two refrigerants (R-12 and R-22) were considered as working fluids. The representation of the thermodynamic properties of these pure components is described in the following section.

WORKING FLUID THERMODYNAMIC PROPERTIES

ASPEN offers the user a number of options for computing thermophysical properties. Therefore, a preliminary evaluation of the ability of the available equations of state, as implemented in their respective option sets in ASPEN, was performed to define the most accurate option set for use in this study.

A substantial number (74) of data values for isobutane vapor pressure, vapor and liquid densities, and vapor and liquid enthalpy departures in the temperature range 255 to 408 K were available from Starling (1977) and ASHRAE (1969). Consequently, the ASPEN data regression system (DRS) was used to fit model-specific parameters for the Redlich-Kwong-Soave (RKS), the Peng-Robinson (PR), and the conformal solution theory Benedict-Webb-Rubin (BWR) equations of state, as implemented in SYSOP3, SYSOP4, and SYSOP5, respectively. Absolute average relative deviations for calculated results, which are a measure of lack-of-fit, are shown in Table I. Although the BWR results are not uniformly better than the RKS and PR results, the accuracy of the conformal solution theory BWR equation of state contained within SYSOP5 is superior, overall.

TABLE I. ABSOLUTE AVERAGE RELATIVE DEVIATIONS FOR THE REGRESSION OF ISOBUTANE DATA

Property	SYSOP3	SYSOP4	SYSOP5
Vapor pressure, kPa	0.66	0.64	0.25
Liquid enthalpy departure, kJ/kmol	0.88 (-)*	0.41 (-)	0.26
Vapor enthalpy departure, kJ/kmol	19.3 (+)	18.2 (+)	5.1 (+)
Liquid density, kg/m^3	9.3 (-)	3.6 (+)	5.9 (-)
Vapor density, kg/m^3	0.25	0.58 (+)	1.0 (+)

*(-), (+) indicates bias, if present, negative or positive.

Therefore, SYSOP5 was selected for all cycle calculations. It should be noted that DRS allows estimation of only three of the eleven parameters available in the conformal solution theory BWR. Therefore, the results reported here are not necessarily indicative of the capabilities of the conformal solution theory BWR.

The parameters determined for SYSOP5 by DRS were used for all calculations of isobutane thermodynamic properties. With the following exceptions, data from the ASPEN pure-component data bank were used for all calculations of thermodynamic properties for propane, R-12 and R-22.

(1) Parameters for propane were taken from Brulé (1979).

(2) Vapor pressure data for R-12 from ASHRAE (1969) were used with DRS to determine parameters for the extended Antoine equation.

ANALYSIS OF THERMODYNAMIC RESULTS

Four thermodynamic cycles were considered with isobutane, propane, R-12, and R-22, respectively, as working fluids. None of the systems utilized superheat, and only the isobutane system utilized regeneration. (These choices of superheat and regeneration were predetermined based on earlier screening calculations not reported here.)

Table II presents pertinent performance parameters for the four systems. The net cycle power and thermal efficiencies are essentially equivalent for the four cycles, considering the inaccuracies inherent in the calculations, although the best balance of net power and thermal efficiency was obtained with isobutane. However, one might elect to use R-22 to eliminate the investment cost of fire prevention measures. Other considerations would include the higher system pressure for R-22, which would increase the equipment costs. Sorting out some of these considerations is possible using ASPEN's cost blocks and the economic analysis package.

TABLE II. PERFORMANCE OF POWER RECOVERY SYSTEMS

Fluid	Net Power (MW)	Thermal Efficiency (percent)
Isobutane	9.3	7.8
Propane	9.1	7.5
R-12	9.0	7.7
R-22	9.1	7.8

RESULTS OF ECONOMIC EVALUATION

A preliminary economic evaluation of the four cycles shown was completed, using investment per kilowatt as a criterion. A complete economic evaluation was not performed because of constraints of time. The costs for all the components used, except for the turbine, were estimated using ASPEN. Turbine costs were obtained separately, since ASPEN does not provide turbine costs. For the same four systems described above, Table III presents the purchased equipment costs. Note that the cost of

TABLE III. PURCHASED EQUIPMENT COST PER KILOWATT

Fluid	$/KW
Isobutane	1900
Propane	1700
R-12	1400
R-22	1700

a fire protection system is not included for the two hydrocarbons, and that their investment costs would necessarily be higher than shown. Based on purchased equipment per kilowatt, the clear winner is R-12. Referring to Table II, we see that selecting R-12 loses very little in terms of thermal efficiency or net power.

This study results in the following conclusions:

(1) ASPEN is a useful tool for thermodynamically evaluating the performance of Rankine cycles for power production.

(2) ASPEN facilitates the economic evaluation of thermodynamically promising systems.

(3) Thermodynamic performance alone, whether based on elementary first law considerations or more sophisticated second law considerations, does not necessarily lead to an economically optimum system design.

REFERENCES

ASHRAE, ASHRAE Thermodynamic Properties of Refrigerants (1969).

Brulé, M. R., Lee, L. L., Starling, K. E., Predicting Thermodynamic Properties for Fossil Fuel Chemicals, Chemical Engineering, 86 (23), 155 (1979).

Starling, K. E., et. al., Self-Consistent Correlation of Thermodynamic and Transport Properties, Oklahoma State University Report No. GRI-AGA/BR-111-1/77-36 (1977).

Work supported by the U. S. Department of Energy under DOE Contract DE-AC07-76ID0570.

COMPARISONS OF DISTILLATION NETWORKS: EXTENSIVELY STATE OPTIMIZED VS. EXTENSIVELY ENERGY INTEGRATED

P.A. MINDERMAN, JR.
and
D.W. TEDDER

School of Chemical Engineering
Georgia Institute of Technology
Atlanta, Georgia 30332

INTRODUCTION

The synthesis of a low-cost process flowsheet for a separation system has become an evermore important task in the engineering world. Due to increasing energy costs, more emphasis is being placed on system optimization in the minimization of flowsheet cost. To achieve this objective, two procedures may be considered. In the first case,

the network may be synthesized with the emphasis on energy integration (see Sophos et al, (1978) or Morari and Faith, (1980) for example). Individual state variables are adjusted for desired integration schemes. Another option is to extensively optimize individual separators and then apply energy integration algorithms to the resultant

---

P. A. Minderman, Jr. is now with Tennessee Eastman Company, Kingsport, Tennessee

flowsheets (see Minderman, (1981) for example). The difference in the procedures is the set of variables which is weighted more heavily. In this work, comparisons will be made between published energy integrated systems and extensively state optimized systems generated by a simple synthesis scheme.

Previous Work on the Synthesis of Separation Sequences

In the synthesis of separation systems, the importance of configuration has been known for years. Thormann (1928) first recognized the problem although did not propose any solutions to it. Today efforts can be classified into three basic groups: heuristic synthesis, evolutionary synthesis, and algorithmic synthesis. As this work considers both simple and complex fractionators (Tedder, (1975)), emphasis is placed on those procedures which consider both possibilities.

To date, complex fractionators have only been considered in heuristic studies. The size of the solution space [Tedder, (1975)] quickly makes many algorithmic procedures, e.g. Hendry and Hughes (1972) and Westerberg and Stephanopoulos (1975), unmanagable. Evolutionary procedures such as that of Seader and Westerberg (1977) are certainly efficient; however, they do not guarantee optimality as the evolutionary rules are generally simple relaxations of heuristics. Hence, only those works which include complex fractionators are summarized here. For complete reviews of separation synthesis, the reader is referred elsewhere [Hlavacek, (1978) and Minderman, (1981)].

Complex fractionators were first considered by Grunberg (1960); however, Petlyuk et al (1965) were the first to compare materially coupled configurations (Figures 3 and 5) with the direct and inverted sequences (Figure 1). In examining ten ternary feeds, the coupled designs required, in general, less vapor than the simple configurations. These results were confirmed by Stupin (1970) in a study which also included a thermally coupled configuration which is not considered here.

In an encompassing study of both simple and complex fractionators, Tedder (1975) developed heuristics for the eight designs illustrated in Figures 1-5. The study used the shortcut models summarized later in this paper. Forty-two ternary feeds were examined at two different sets of utility costs. Designs were extensively state-optimized based on venture cost [Rudd and Watson, (1968)]. The heuristics were expressed as a function of the Ease of Separation Index (ESI) defined as:

$$ESI = K_a K_c / K_b^2$$

where K_a = distribution coefficient of overheads product

K_b = distribution coefficient of middle product

K_c = distribution coefficient of bottoms product

Using this parameter, the heuristics may be paraphrased as follows: (all compositions are in mole percent)

For ESI less than 1.6:

1. If 40 to 80% of the feed is middle product and the amounts of the overheads and bottoms products are nearly equal, then favor Design V.
2. If more than 50% of the feed is middle product and less than 5% is bottoms product, then favor Design VI.
3. If more than 50% of the feed is middle product and less than 5% is overheads product, then favor Design VII.
4. If less than 15% of the feed is middle product and the amounts of the overheads and bottoms products are nearly equal, then favor Design III.
5. Otherwise, favor Design I or II whichever removes the most plentiful component first.

For ESI greater than or equal to 1.6:

1. If more than 50% of the feed is bottoms product, then favor Design II.
2. If more than 50% of the feed is middle product and from 5 to 20% is bottoms product, then favor Design V.
3. If more than 50% of the feed is middle product and less than 5% is bottoms product, then favor Design VI.
4. If more than 50% of the feed is middle product and less than 5% is overheads product, then favor Design VII.
5. Otherwise, favor Design III.

Doukas and Luyben (1978) completed a similar study to that of Tedder (1975). Considering Designs I, II, V, VI, and VII for ternary feeds of benzene, toluene, and o-xylene, heuristics were developed using an equilibrium stage model. Optimization was performed using the Fenske-Underwood-Gilliland shortcut methods for simple towers and was empirically performed for the sidestream towers. Results are similar to those of Tedder:

At the set of relative volatilities (6.7, 2.4, 1) (ESI = 1.2):

1. For a feed of less than or equal to 10% overheads, favor Design VII.

2. For a feed of greater than 10% overheads but less than 30%, favor Design V.

As the relative volatilities decreased, the bound of optimality between VII and V also decreased.

3. For a feed of less than or equal to 10% bottoms, favor Design VI.
4. For a feed of greater than 10% bottoms, favor Design V.

At lower relative volatilities (3, 2, 1) (ESI = 0.75), the direct design was found to be optimal at bottoms concentrations of 10 and 20%.

Freshwater et al (1981) also studied ternary feeds, however only considered simple configurations. Furthermore, they studied the Ease of Separation Index of Tedder (1975). For an ESI less than unity, results agreed with those of Tedder. However for ESI greater than 1.6, the results conflicted with those of Tedder when the difficult separation was between the middle and bottoms components. Tedder attributes these differences to the fact that Freshwater et al did not optimize their sequences with respect to operating pressure.

Problem Definition

Given a feed stream at a known state, systematically synthesize a process that isolates the specified products at a minimum venture cost.

To simplify the solution space for the synthesis procedure, certain assumptions will be made. First only straight distillation will be considered although both simple and complex fractionators will be examined. The seven configurations of interest are given in Figures 1-4, and the corresponding design variables are given in Table 1. Secondly, sloppy splits of the key components (i.e. Figure 5) will not be allowed. Justification of this assumption is based on the work of Tedder (1975) in which sloppy splits were not found to be optimal in the separation of ternary feeds. Additional restrictions include:

1. All tower products are saturated liquids.
2. Only total condensers and reboilers are used.
3. The cost of changing pressure of streams between columns is negligible.

The last assumption is also based on the work of Tedder (1975).

Table 1. Design Variables

Design	Variables
I, II, VI, VII	1. Overheads pressure 2. Fraction vaporization of the feed (f1) 3. Actual vapor to minimum vapor ratio (V/Vmin1)
III, IV	1. Overheads pressure 2. Fraction vaporization of the feed (f1) 3. Actual vapor to minimum vapor ratio, upstream tower (V/Vmin1) 4. Actual vapor to minimum vapor ratio, downstream tower (V/Vmin2)
V	1. Overheads pressure 2. Fraction vaporization of the feed, upstream tower (f1) 3. Actual vapor to minimum vapor ratio, upstream tower (V/Vmin1) 4. Degree of separation, upstream tower (Stages) 5. Overheads to feed ratio, upstream tower (P1/F) 6. Fraction vaporization of upper feed, downstream tower (f2) 7. Fraction vaporization of lower feed, downstream tower (f3) 8. Actual vapor to minimum vapor ratio, downstream tower (V/Vmin2)

SYNTHESIS PROCEDURE

The distillation trains are synthesized by first generating a set of low cost flowsheets using modified and extended heuristics of Tedder (1975) and then applying a rigorous branch and bound search (Phillips et al, 1976) over the unintegrated solution space using the heuristic generated flowsheet cost as an upper bound to the optimal cost. The resultant flowsheets are then energy integrated using the method of Umeda et al (1979). In applying the heuristics of Tedder (1975), the heuristics will be extended to feeds of more than three components. In this case, a psuedo-ternary feed is defined such that the relative volatilities are those of the three key components; and the psuedo-ternary composition are those of the overheads key and lighter components, the bottoms key and heavier components, and remaining components. Then, given the ternary or psuedo-ternary feed stream, the heuristics are applied as presented in Tedder and Rudd (1978).

In generating heuristic flowsheets, the heuristics will not only be applied to the original feed but also to all intermediate feeds of three or more components. Since heuristic decisions are not always clearly defined, the above procedure generates a set of flowsheets from which a near-optimal flowsheet may be selected. Cases where heuristics conflict can be avoided.

The rigorous branch and bound search is started using the least cost heuristic flowsheet as generated from the above procedure as the upper bound to the optimal flowsheet cost. The tree is searched until a branch either exceeds the current upper bound or a lower cost flowsheet is generated. In the latter case, the upper bound is then updated. In either case, the search is continued until the entire tree is fathomed.

Optimization and Rating of Individual Separators

As in the work of Tedder (1975), individual separators will be represented and evaluated by shortcut methods. First, the material balance is specified by a species allocation method where the composition of the adjacent non-key components are fixed. A design is then represented by a set of nonequilibrium nodes where the countercurrent vapor and liquid streams are considered to be at the same composition. Having specified the composition nodes, the only remaining trial-and-error calculations are for estimating feed tray composition for product streams resulting from a sloppy split (i.e. the upstream tower in Design V) and for bubble and dew point calculations. Furthermore, with the composition nodes specified, the interactions between individual designs due to composition changes do not occur. Hence the operating state of an individual design can be optimized very efficiently.

For the details of these calculations, the reader is referred to Tedder (1975) or Tedder and Rudd (1978) for discussion of the species allocation method. As for the optimization of individual designs, the method may be summarized as follows:

1. Assume process configuration.
2. Specify operating state vector.
3. Calculate vapor and liquid temperature profiles.
4. Calculate minimum theoretical stages in each section from the Fenske equation.
5. Calculate minimum vapor rate at the pinch point from the Underwood equation.
6. Calculate the vapor rate at the pinch point by optimization.
7. Calculate the number of theoretical stages in each section from the Gilliland correlation.
8. Calculate the actual number of stages by multiplying the number of theoretical stages by the stage efficiency and adding 10% overdesign.
9. Calculate the vapor and liquid flow rate profiles.
10. Size all equipment items.
11. Choose the cheapest utilities from those satisfying temperature approach.
12. Calculate the capital investment and annual operating cost.
13. Choose a new operating state to minimize venture cost.

For detailed discussion of this procedure, the reader is referred to Whittlesey (1981).

The validity of this design method is not addressed in this work; it has been illustrated that the model is of sufficient accuracy for a screening study. For discussions on the accuracy of the design method, the reader is referred to Tedder (1975) and Whittlesey (1981). Of particular interest to this work, Whittlesey compared several fractionators generated in this work with the corresponding towers rated by the FRAKB equilibrium stage model from the FLOWTRAN steady state simulation system (Seader et al, 1977).

Tables 2 through 5 summarize the economic data and equations and physical properties used in the rating model. The capital cost equations in Table 3 were regressed from information in Guthrie (1970). The utility data in Table 4 were based on information given in Aries and Newton (1955) after appropriate scaling. Finally the physical properties are given in Table 5. All of these properties except the Cavett vapor enthalpy regressions (Cavett, 1962) were taken from Reid and Sherwood (1966).

Table 2. Summary of Design Data

Venture Cost = 0.52 (Annual operating cost) + 0.2996 (Capital investment)

Income tax rate: 48% of gross earnings
Working capital: 18% of fixed capital investment
Project life: 20 years
Average risk factor: 8%
Minimum rate of return: 20%
Taxes: 4% of fixed capital investment
Insurance: 1% of fixed capital investment
Maintenance: 5% of fixed capital investment
Annual operating time: 8500 hours
Compressor efficiency: 80%
Materials of construction: carbon steel

Table 4. Range of Utility Costs

Utility	T (K)	Cost
Steam ($/GJ)	626	16.63
	554	12.82
	504	10.26
	459	7.74
	418	5.72
	388	4.18
Cooling Water (¢/m^3)	297	6.60
Ammonia ($/GJ)	274	16.49
	255	29.67
	246	39.62
	232	49.44
	197	77.74
Electricity ($/kJ)	-	9.72

Table 3. Capital Cost Estimation

Heat exchanger capital investment:
$CAP = b \, (area / 64.42)^{0.65}$
where $b = 23730$ for $P < 1.034$ MPa
       $= 23730 \, [1 + 0.145(P - 1.034)]$ for $P \geq 1.034$ MPa

Vertical pressure vessel capital investment:
$CAP = b \, (D/c)^{1.04694}$
where $b = 8680$ for $P < 0.345$ MPa
       $= 8680 \, [1 + 0.145(P - 0.345)]$ for $P \geq 0.345$ MPa
     $c = 6.360/h + 0.2371$ for $h \leq 4.572$ m
       $= 10.099/h + 0.0387$ for $h \geq 4.572$ m
     D = tower diameter (m)
     h = tower height (m)
     P = operating pressure on top tray of tower (MPa)

Tray capital investment:
$CAP = 1.4 \, (c)(n - 3.048)$
where $c = 2.64 \, D - 0.3712$ for $D < 1.22$ m
       $= 5.08 \, D - 10.16$ for $1.22 \leq D \leq 1.83$ m
       $= 7.11 \, D - 22.34$ for $1.83 \leq D \leq 2.13$ m
       $= 7.62 \, D - 22.86$ for $D < 2.13$ m

Horizontal pressure vessel capital investment:
$CAP = b \, (D/c)^{0.97468}$
where $b = 4523.75$ for $P < 0.345$ MPa
       $= 4523.75 \, [1 + 0.145(P - 0.345)]$ for $P \geq 0.345$ MPa
     $c = 2.64 \, D - 0.3712$ for $D \leq 1.22$ m
       $= 5.08 \, D - 10.16$ for $1.22 \leq D \leq 1.83$ m
       $= 7.11 \, D - 22.34$ for $1.83 \leq D \leq 2.13$ m
       $= 7.62 \, D - 22.86$ for $D > 2.13$ m

Compressor capital investment:
$CAP = 2718.207 \, (HP/745.7)^{0.73532}$
where HP = theoretical power required (W)

Table 5.   Summary of Physical Properties

Vapor-Liquid Equilibria
    Chao-Seader correlation
        Redlich-Kwong vapor fugacity coefficient
        Scatchard-Hildebrand liquid activity coefficient
        Chao-Seader standard-state liquid fugacity
Vapor enthalpies
    Cavett polynomial correlations
Latent heats of vaporization
    Watson correlation
Liquid enthalpies
    H liquid = H vapor - H vaporization
Vapor densities
    Redlich-Kwong equation of state
Liquid densities
    Cavett estimation
Liquid viscosities
    Andrade equation

Mixture properties (densities, enthalpies, and viscosities) are evaluated by a molar averaging of the pure component properties at the node of interest.

A SEPARATION OF LIGHT PARAFFINS

In order to illustrate the procedure as outlined above, an example has been chosen which has appeared in at least seven prior investigations; thus, it should be a good candidate for comparison.

Example Description

This example is usually attributed to Rathore et al (1974 a and b); however, it was reported previously by Heaven (1969). The feed stream contains five light ideal hydrocarbons at the following composition:

Component	Mole Fraction
Propane (A)	0.05
i-Butane (B)	0.15
n-Butane (C)	0.25
i-Pentane (D)	0.20
n-Pentane (E)	0.35

The feed rate is 907 kmol/hr, and the feed pressure is atmospheric. The recovery specifications for all components were 98% in the original work and in many succeeding investigations.

All previous investigators have restricted their work to only simple configurations. For such a solution space, there are 14 distinguishable flowsheets for the five-component system (Heaven, 1969). By adding complex fractionators, this number increases dramatically to 194; however, the minimum number of subproblems which must be optimized to consider all designs is much lower, 95.

The species allocation method was used to calculate the compositions and flow rates of all intermediate and final products. A purity specification of 1% was assumed for the adjacent non-key components. The results of this method are shown in Table 6. As shown, the recovery specifications of the key components in the product streams are very nearly 98%. In order to enable comparisons between this work and those of previous workers, previous results have been rated using the species allocation results and the economics given previously. The state variables were either fixed or optimized according to the results reported by the workers and are summarized in Table 7.

Previous Results

Heaven (1969) first examined this example in an early study. The fraction vaporization of the feed was set to zero, and the overhead pressures and reflux ratios were optimized. However, these results were not reported. Therefore, his sequence was optimized using the Composition Node Design Method. The resultant flowsheet is shown in Figure 6.

Rathore et al (1974 a and b) studied several cases of this example. For the case (1974 b) in which they optimized the overheads pressure and reflux ratio, but not the fraction vaporization of the feed and used their "more realistic" utility cost equations, the best flowsheet without integration is shown in Figure 7. The best flowsheet after energy integration is given in Figure 8. Work by other investigators, Freshwater and Ziogou (1976) and Gomez and Seader (1976), also illustrated this example, but did not generate any different flowsheets than these of Rathore et al.

The next workers who generated an improved flowsheet were Sophos et al (1978) by incorporating more heat integration than previous flowsheets. Their result is shown in Figure 9. As stated previously, operating variables were set in order to get a good energy match.

Table 6. Stream Rates and Compositions from Species Allocation Method

--Original Feed Stream--

Flow Rate (kmol/hr)	Composition (mole fraction)				
	Propane	i-Butane	n-Butane	i-Pentane	n-Pentane
907.18	0.05000	0.15000	0.25000	0.20000	0.35000

--Product Streams--

Flow Rate (kmol/hr)	Composition (mole fraction)				
	Propane	i-Butane	n-Butane	i-Pentane	n-Pentane
44.54	0.98573	0.01000	0.00425	9.56E-6	2.86E-6
135.62	0.01000	0.97994	0.01000	4.68E-5	1.74E-5
230.01	0.00040	0.01000	0.97149	0.01000	0.00811
180.00	3.83E-5	0.00241	0.01000	0.97756	0.01000
317.02	5.5E-16	4.7E-10	3.66E-8	0.01000	0.99000

--Intermediate Feed Streams--

Flow Rate (kmol/hr)	Composition (mole fraction)				
	Propane	i-Butane	n-Butane	i-Pentane	n-Pentane
180.16	0.25122	0.74015	0.00858	3.76E-5	1.38E-5
365.62	0.00396	0.36977	0.61485	0.00631	0.00511
410.00	0.00024	0.00666	0.54938	0.43477	0.00894
497.00	1.39E-5	0.00087	0.00362	0.36041	0.63508
410.18	0.11057	0.33071	0.54855	0.00562	0.00455
545.62	0.00267	0.24858	0.41531	0.32672	0.00672
727.02	0.00014	0.00376	0.30982	0.24955	0.43673
590.16	0.07686	0.23057	0.38429	0.30206	0.00621
862.64	0.00169	0.15723	0.26269	0.21033	0.36807

Umeda et al (1979) applied their heat integration procedure to the flowsheets of Rathore et al and obtained significantly improved economics although they did not report any economic evaluation in their presentation. In their article, they only reported the temperatures and heat loads of the condensers and reboilers. For comparison, the tower pressures were fixed to give the reported temperatures, and the vapor to minimum vapor ratios and fraction vaporizations of the feeds were optimized. The heat loads were not examined due to lack of knowledge of the enthalpy correlations used by the investigators. Their results are presented in Figures 10 and 11.

Recently Morari and Faith (1980) also presented alternative flowsheets for this example. The towers were rated using the operating pressures and reflux ratios given in Faith (1979). As in that work, the fraction vaporizations of the feeds were set to zero. Two cases were considered; one in which steam splitting for energy integration was permitted and one in which it was not. The modeling of their results is shown in Figures 12 and 13.

Flowsheet Generation

The heuristics were applied to all feed streams as previously discussed. The four flowsheets generated by the heuristics are presented in Figure 8; all designs are simple columns. The lowest cost flowsheet is shown in Figure 14. Next, the branch and bound search was performed with the heuristic generated cost of $3.342 million as the initial upper bound. The upper bound was improved four times over the tree search until the optimal flowsheet before integration was obtained. This flowsheet is shown in Figure 15. Since this flowsheet cost ($3.230 million) was so close (4%) to the heuristic flowsheet, the tree was re-examined to produce all flowsheets of a lower cost than the heuristic one. These are summarized in Table 9. Because of the size of the branch and bound tree, it is not presented here; but rather the interested reader is referred to Minderman (1981).

Table 7. State Variables for Previously Reported Flowsheets

Split	Pressure (MPa)	Feed Fraction Vapor	$\overline{V/V_{min}}$	Unintegrated Tower Venture Cost (MM$)
Flowsheet -- Heaven				
ABC/DE	0.477	0.000	1.025	0.907
AB/C	0.620	0.000	1.020	1.191
A/B	1.296	0.000	1.065	0.171
D/E	0.165	0.000	1.047	1.532
Flowsheet -- Rathore et al (without energy integration)				
ABC/DE	0.967	0.000	1.002	1.201
AB/C	0.847	0.000	1.026	1.199
A/B	1.536	0.000	1.026	0.169
D/E	0.285	0.000	1.035	1.636
Flowsheet -- Rathore et al (with energy integration)				
ABC/DE	0.967	0.000	1.022	1.201
A/BC	1.536	0.000	1.031	0.295
B/C	0.922	0.000	1.027	1.182
D/E	0.681	0.000	1.021	2.357
Flowsheet -- Sophos et al				
ABC/DE	0.552	0.000	1.056	0.939
A/BC	1.462	0.000	1.090	0.332
B/C	0.790	0.000	1.093	1.214
D/E	0.831	0.000	1.095	2.679
Flowsheet -- Umeda et al (Case I)				
ABC/DE	0.668	-0.161	1.050	0.971
A/BC	1.508	-0.500	1.029	0.269
B/C	0.552	0.100	1.050	1.149
D/E (55%)	0.758	-0.100	1.030	1.403
D/E (45%)	0.400	0.193	1.027	0.835
Flowsheet -- Umeda et al (Case II)				
ABC/DE	0.889	-0.135	1.020	1.177
A/BC	1.538	-0.352	1.048	0.266
B/C	0.703	0.300	1.024	1.137
D/E	0.745	0.081	1.021	2.407
Flowsheet -- Morari and Faith (without stream splitting)				
A/BCDE	2.756	0.000	1.032	1.721
BC/DE	1.074	0.000	1.039	0.989
B/C	0.679	0.000	1.047	1.157
D/E	0.203	0.000	1.059	1.602
Flowsheet -- Morari and Faith (with stream splitting)				
ABC/DE	1.165	0.000	1.042	1.272
A/BC	1.955	0.000	1.027	0.322
B/C	1.165	0.000	1.109	1.294
D/E	0.212	0.000	1.059	1.614

Table 8. Flowsheets Generated by Heuristics Approach

Split	Pressure (MPa)	Feed Fraction Vapor	$\overline{V/V_{min}}$	Unintegrated Tower Venture Cost (MM$)
Flowsheet -- 1				
ABC/DE	0.502	-0.274	1.006	0.892
A/BC	1.344	-0.500	1.050	0.268
B/C	0.538	0.354	1.022	1.113
D/E	0.165	0.377	1.047	1.497
				3.700
Flowsheet -- 2				
A/BCDE	1.317	-0.329	1.023	0.761
BCD/E	0.282	0.576	1.078	1.782
B/CD	0.607	0.000	1.031	1.296
C/D	0.447	-0.183	1.023	0.394
				4.233
Flowsheet -- 3				
ABCD/E	0.552	-0.437	1.033	1.396
ABC/D	0.534	-0.002	1.065	0.565
A/BC	1.344	-0.500	1.050	0.268
B/C	0.538	0.354	1.022	1.113
				3.342
Flowsheet -- 4				
AB/CDE	0.614	-0.529	1.023	1.561
A/B	1.524	-0.529	1.065	0.161
CD/E	0.243	0.192	1.081	1.671
C/D	0.447	-0.183	1.023	0.394
				3.787

Energy Integration of Resultant Flowsheets

The energy integration procedure was applied to many of the flowsheets listed in Table 9, and the results can be classified into two categories: integration of complex configurations and integration of simple configurations.

Integration of complex configurations in this five-component example was not found to be economically attractive in all cases. Changing the operating pressure away from optimal always more than offset the savings due to exchanger matching. To illustrate this observation, the complex fractionator in the best complex flowsheet (Figure 15) was studied parametrically by fixing the operating pressure and optimizing over the remaining three variables. The results showing the product bubble points and steam requirements as functions of pressure are shown in Table 10. The resultant venture costs are plotted versus pressure and shown in Figure 17.

Table 9. Summary of Flowsheets Cheaper than Heuristics

Design Number	Split	Tower Venture Cost (millions of $)
IV	ABC/D/E	1.885
I	AB/C	1.184
I	A/B	0.161
		3.230
IV	ABC/D/E	1.885
I	A/BC	0.268
I	B/C	1.113
		3.266
I	ABCD/E	1.396
V	AB/C/D	1.720
I	A/B	0.161
		3.277
I	ABCD/E	1.396
I	A/BCD	0.354
V	B/C/D	1.527
		3.278
I	ABCD/E	1.396
III	AB/C/D	1.740
I	A/B	0.161
		3.297
I	ABCD/E	1.396
IV	AB/C/D	1.744
I	A/B	0.161
		3.301
III	ABC/D/E	1.960
I	AB/C	1.184
I	A/B	0.161
		3.305
I	ABCD/E	1.396
I	ABC/D	0.565
I	AB/C	1.184
I	A/B	0.161
		3.306
III	ABC/D/E	1.960
I	A/BC	0.268
I	B/C	1.113
		3.341

Raising or lowering operating pressures so as to achieve heat exchanger matching was never found to be economical. As shown in Figure 17, a decrease in tower pressure results in substantial increases in cost. Raising pressure to match a condenser from the complex fractionator required very expensive steam for the reboilers because of the large temperature difference across the tower. Thus one was led to look for other possibilities of energy matching.

Vapor recompression or reboiler flashing was the next alternative considered. One can see from Table 10 that the design does not meet Null's (1976) heuristics for favoring these operations. At most pressures:

1. The overheads temperatures were greater than 311 K.
2. The temperature differences between closest products were greater than 19 K.

Multieffect systems (Tyreus and Luyben, 1975) were also considered, especially since the reboiler loads were so large. However, the large temperature difference across the tower always forced the use of an expensive utility at elevated or depressed pressures. Perhaps in larger systems, energy integration with complex fractionators might become a viable alternative.

Integration of simple design flowsheets has been proven to be profitable in many previously discussed works. The integration of the least cost flowsheet before integration is shown in Figure 18. As illustrated, most of the matching occurs between the simple towers; the complex design is only minimally involved. The least cost flowsheet after integration is shown in Figure 19. This flowsheet resulted from the least cost simple flowsheet from the branch and bound search. The venture costs from previous investigations and from this work are ranked according to venture cost and given in Table 11.

As Rathore et al (1974 a and b) discussed, the unintegrated flowsheets have been shown to be upper bounds to the energy integrated ones. However, the conclusion that differs from many preceeding investigators is that the distillation optimization is far more important than previously reported. In the best case of energy integration (Figure 19), the additional cost savings through integration is only 9%. Admittedly only one system has been examined in this work; however, by using high utility costs, the motivation for energy integration should have been greater

Table 10. Parametric Study of Complex Fractionator in Figure 15

Pressure (MPa)	Product Bubble Pts (K)			Heat Loads (GJ/hr)			Steam Requirements (kg/hr)					
	P1	P2	P3	P1	P2	P3	0.172 MPa steam		0.414 MPa steam		1.138 MPa steam	
							P2	P3	P2	P3	P2	P3
0.172	274	325	335	58.6	11.3	50.3	5097	22700	0	0	0	0
0.345	297	348	357	62.9	13.5	57.3	6080	25900	0	0	0	0
0.517	312	363	372	62.3	15.3	57.5	6920	25970	0	0	0	0
0.579	316	368	377	51.5	14.7	47.2	6610	21300	0	0	0	0
0.689	323	375	384	67.8	15.1	62.9	6800	0	0	29530	0	0
0.862	333	385	394	55.6	14.8	51.9	0	0	6930	24390	0	0
1.034	341	394	403	57.6	15.1	53.9	0	0	7060	25340	0	0
1.379	354	409	417	68.4	15.0	65.0	0	0	7040	0	0	32590
1.724	365	420	429	67.8	14.6	65.5	0	0	0	0	7330	32880

where P1 = overheads product
P2 = middle product
P3 = bottoms product

than in previous works. Yet by optimizing more individual configuration variables than previous investigators, energy integration has become less important. Hence the control of the sequences should be simpler.

Table 11. Summary of Reported Results for Light Paraffins Example

Investigator(s)	Venture Cost (million of $)
This work -- optimal simple flowsheet after energy integration	3.016
Umeda et al -- Case I	3.109
This work -- optimal complex flowsheet after energy integration	3.160
This work -- optimal flowsheet before integration	3.230
Morari and Faith -- with stream splitting	3.303
This work -- optimal simple flowsheet before integration	3.306
This work -- heuristic flowsheet	3.342
Sophos et al	3.403
Umeda et al -- Case II	3.412
Heaven	3.801
Rathore et al -- with integration	4.127
Rathore et al -- before integration	4.204
Morari and Faith -- without stream splitting	4.608

SENSITIVITY OF THE MODEL

At first glance, the results of this work appear to contradict other reported efforts in recent literature, i.e., the energy integration was much less significant than the unintegrated state optimization. The authors believe several points help explain these differences.

Firstly, the example presented in this document was a design of a new facility not a study of an existing one. Energy integration will result in more significant savings in a facility where the size of the separators is already fixed.

Secondly, it should be noted that the column and energy interchange synthesis in this effort was sequential rather than simultaneous. We have not directly compared these two synthesis techniques using the same objective function. We have presented the results of this sequential synthesis and the results from a simultaneous one [Morari and Faith, (1980)]. However, the models used in this work and by Morari and Faith were significantly different, thus such a comparison is not a true test between sequential and simultaneous synthesis.

Finally, the previous point raises the issue of the sensitivity of the model in this work to economic changes, to physical properties, and to configuration modeling. As for economic sensitivity, a previous study [Tedder and Rudd, (1978)] optimized three component separations using two different sets of utility costs, where the second set was an order of magnitude larger than the first set. In that study, it was reported that these changes did not have a significant effect on the optimal sequence. Consequently, we did not examine this issue in any detail. Perhaps this is an area where additional effort is needed.

As for physical properties, the models in this work are different than those in most previous works. For example, this work uses the Watson equation for heat of vaporization; most previous works used a cubic polynomial. Again, we have not directly compared various methods on a large-scale problem. For a particular study involving C_2 and C_3 hydrocarbons, we did find that the Watson equation more accurately represented heat of vaporization data than a cubic. Consequently, we did not study the sensitivity of the model to the differing property models.

As for configuration modeling sensitivity, the Composition Node Design Model does not assume constant molal overflow nor constant relative volatility like many previous models. Yet it is not as accurate as say the equilibrium stage models used by Freshwater and co-workers in a single function evaluation. We have not directly studied the sensitivity of optimal flowsheet to the configuration models. More work in these areas is needed.

CONCLUSIONS

Extending known heuristics to the synthesis of large distillation systems with heat integration has been shown to be very successful. The proposed heuristic procedure of generating an upper bound has been observed to generate an excellent bound. Given such a tight bound, the need for a rigorous tree search over the unintegrated solution space has been shown to be superfluous. Evolutionary procedures certainly would have sufficed.

The optimization of the unintegrated distillation designs has been seen to be as important in terms of cost savings as complex energy integrations. By comparing an extensively optimized distillation sequence with published extensively integrated sequences, the cost savings due to such complex energy integration was seen to be small for new facility designs. While the magnitude of this observation may be due to the particular example considered, one can nevertheless conclude that, in general, investigators do not place enough emphasis on this optimization.

Because of the differences between this work and previous works, additional efforts are needed to study the sensitivity of synthesis models to optimization technique, to the economic function, to the physical properties, to the configuration models, and to feed disturbances.

ACKNOWLEDGMENT

Acknowledgment is made to J. J. Siirola of Tennessee Eastman Company for his helpful critique of this paper.

LITERATURE CITED

Aries, R. S. and R. D. Newton Chemical Engineering Cost Estimation. McGraw-Hill, New York, (1955).

Cavett, R. H. "Physical Data for Distillation Calculations - Vapor-Liquid Equilibria," Twenty-Seventh Midyear Meeting of API Division of Refining, 42, (3), 351, May (1962).

Doukas, N. and W. L. Luyben. "Economics of Alternative Distillation Configurations for the Separation of Ternary Mixtures," Ind Eng Chem Process Des Dev, 17, 3, 272, (1978).

Faith, D. C., III. "The Synthesis of Distillation Trains with Heat Integration," MS Thesis, University of Wisconsin, Madison, (1979).

Freshwater, D. C., B. D. Henry, and W. G. Kirchner. "The Configuration of Multi-Component Distillation Systems," submitted to AIChE J in 1981.

Freshwater, D. C. and E. Ziogou. "Reducing Energy Requirements in Unit Operations," Chem Eng J (Lausanne), 11, 3, 215, (1976).

Gomez, A. and J. D. Seader. "Separation Sequence Synthesis by a Predictor Based Ordered Search," AIChE J, 22, 970, (1976).

Grunberg, J. F. "The Reversible Separation of Multicomponent Mixtures," Advances in Cyrogenic Engineering, 2, New York, (1960).

Guthrie, K. M. "Capital Cost Estimating," Modern Cost Engineering Techniques, 80, McGraw-Hill, New York, (1970).

Heaven, D. L. "Optimum Sequencing of Distillation Columns in Multicomponent Fractionation," MS Thesis, University of California, Berkeley, (1969).

Hendry, J. E. and R. R. Hughes. "Generating Process Flowsheets," Chem Eng Progr, 68, 6, 69, (1972).

Hlavecek, V. "Synthesis in the Design of Chemical Processes," Comp and Chem Eng, 2, 67, (1978)

Minderman, P. A., Jr. "Heuristic Synthesis of Distillation Networks," MS Thesis, Georgia Institute of Technology, Atlanta, (1981).

Morari, M. and D. C. Faith, III. "The Synthesis of Distillation Trains with Heat Integration," AIChE J, 26, 6, 916, (1980).

Null, H. R. "Heat Pumps in Distillation," Chem Eng Progr, 72, 7, 58, (1976).

Petlyuk, F. B., V. M. Platonov, and D. M. Slavinckii. "Thermodynamically Optimal Method for Separating Multicomponent Mixtures," Int Chem Eng, 5, 555, (1965).

Phillips, D. T., A. Ravindran, and J. J. Solberg. Operations Research Principles and Practice. John Wiley and Sons, New York, (1976).

Rathore, R. N. S., K. A. Van Wormer, and G. J. Powers. "Synthesis Strategies for Multicomponent Separation Systems with Energy Integration," AIChE J, 20, 3, 491, (1974 a).

_____. "Synthesis of Distillation Systems with Energy Integration," AIChE J, 20, 5, 940, (1974 b).

Reid, R. C. and T. K. Sherwood. The Properties of Gases and Liquids. McGraw-Hill, New York, (1966).

Rudd, D. F. and C. C. Watson. Strategy of Process Engineering. Wiley, New York, (1968).

Seader, J. D., W. D. Seider, and A. C. Pauls. FLOWTRAN Simulation -- An Introduction, 2nd edition, CACHE, Cambridge, MA (1977).

Seader, J. D. and A. W. Westerberg. "A Combined Heuristic and Evolutionary Strategy for Synthesis of Simple Separation Sequences," AIChE J, 23, 6, 951, (1977).

Sophos, A., G. Stephanopoulos, and M. Morari. "Synthesis of Optimum Distillation Sequences with Heat Integration Schemes," 71st Annual AIChE Meeting, Miami, FL, (1978).

Stupin, W. J. "The Separation of Multicomponent Mixtures in Thermally-Coupled Distillation Systems," Ph D Thesis, University of Southern California, Los Angeles, (1970).

Tedder, D. W. "The Heuristic Synthesis and Topology of Optimal Distillation Networks," Ph D Thesis, University of Wisconsin, Madison, (1975).

Tedder, D. W. and D. F. Rudd. "Parametric Studies in Industrial Distillation," AIChE J, 24, 2, 303, (1978).

Thormann, K. Destillieren und Rektifizieren. Spamer, Leipzig, (1928).

Tyreus, B. D. and W. L. Luyben. "Two Towers Cheaper Than One," Hydrocarbon Processing, 54, 7, 93, (1975).

Umeda, T., K. Niida, and K. Shiroko. "A Thermodynamic Approach to Heat Integration in Distillation Systems," AIChE J, 25, 3, 423, (1979).

Westerberg, A. W. and G. Stephanopoulos. "Studies in Process Synthesis - I. Branch and Bound Strategy with List Techniques for Synthesis of Separation Schemes," Chem Eng Sci, 30, 963, (1975).

Whittlesey III, G. S. "The Design and Optimization of Distillation Columns by the Composition Node Design Method," MS Thesis, Georgia Institute of Technology, Atlanta, (1981).

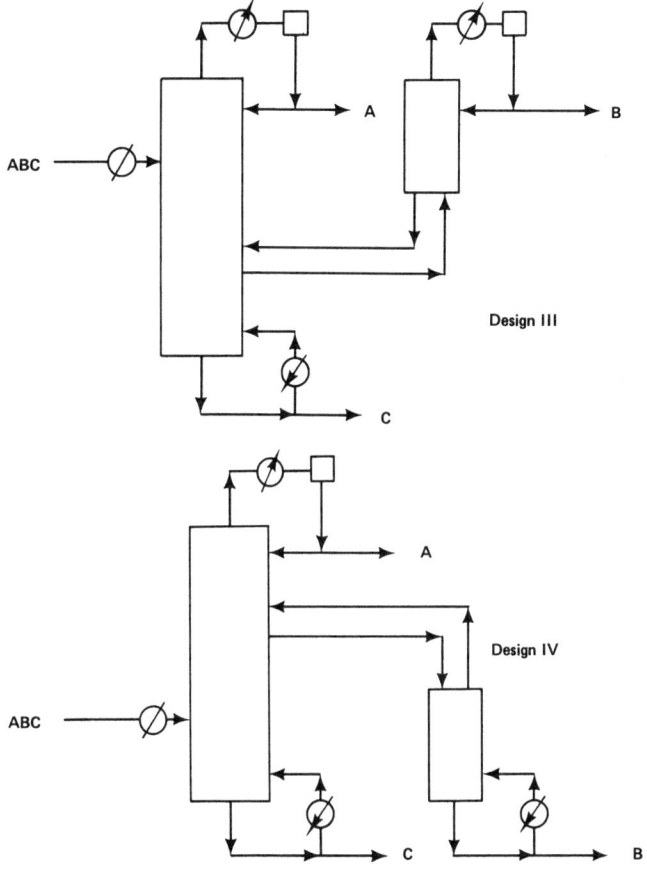

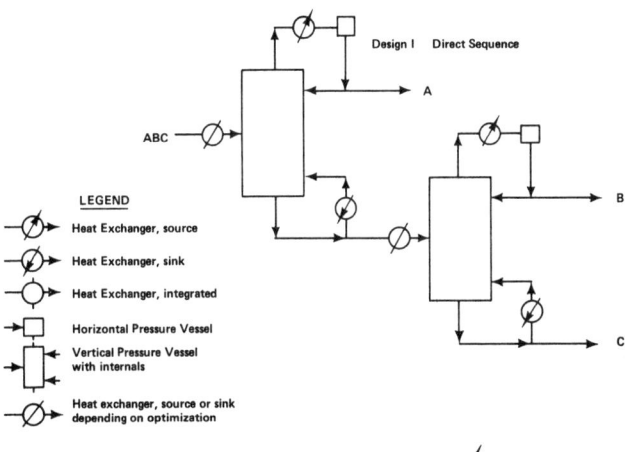

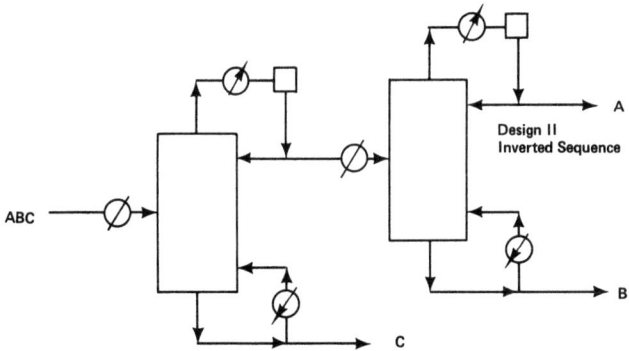

Figure 1. Simple configurations.

Figure 2. Designs III and IV.

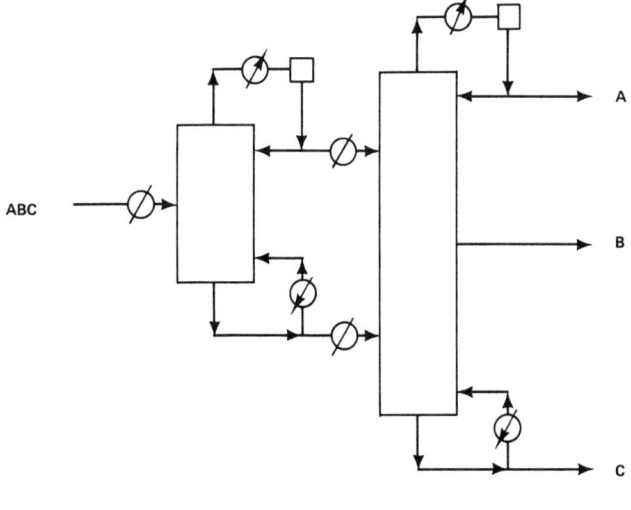

Figure 3. Design V.

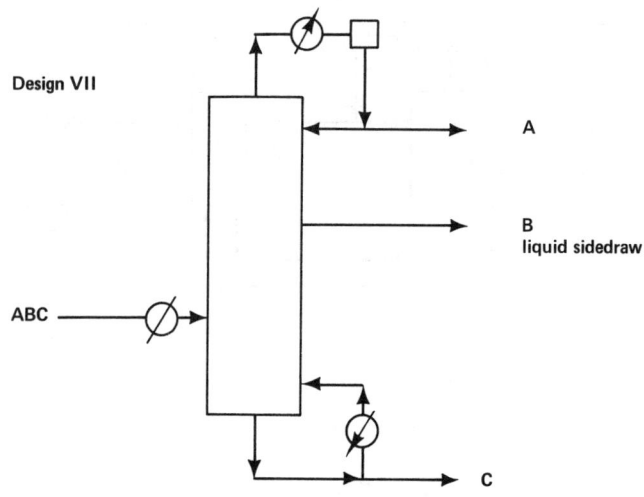

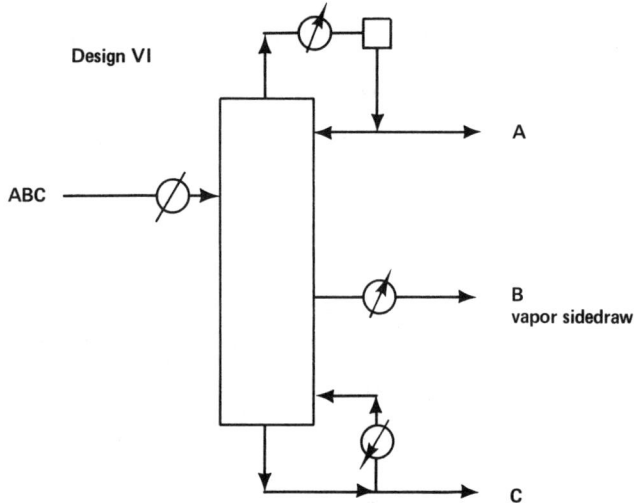

Figure 4. Designs VI and VII.

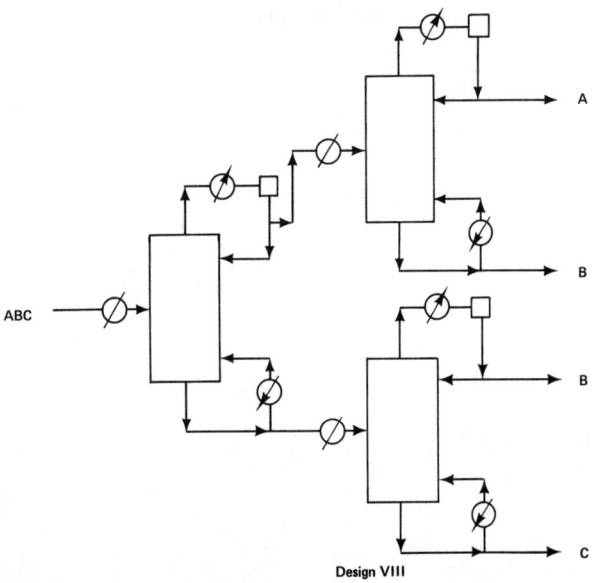

Figure 5. Design VIII.

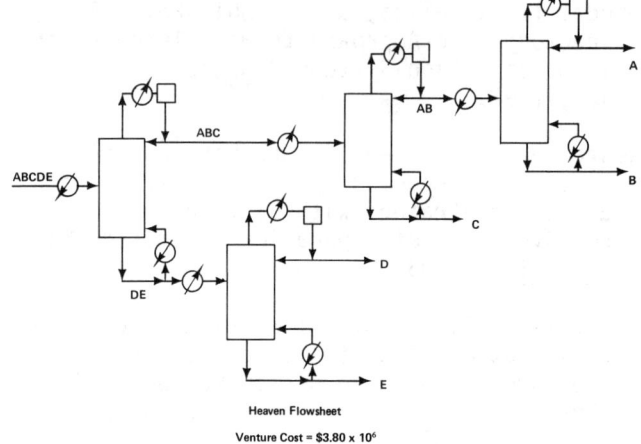

Figure 6. Heaven flowsheet.

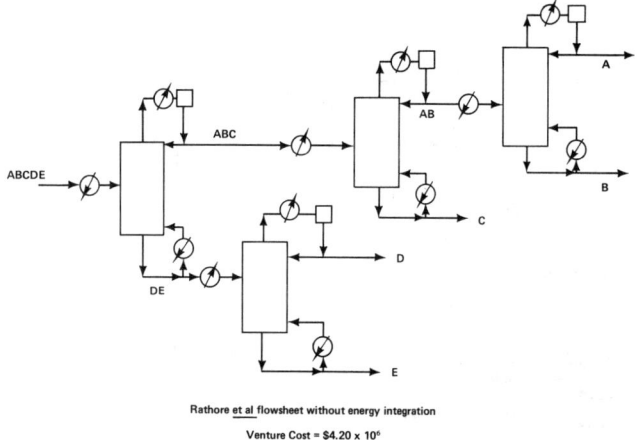

Figure 7. Rathore *et al* flowsheet without energy integration.

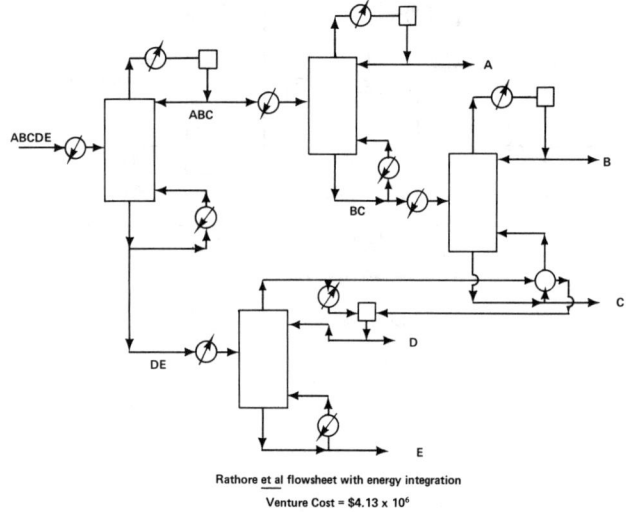

Figure 8. Rathore *et al* flowsheet with energy integration.

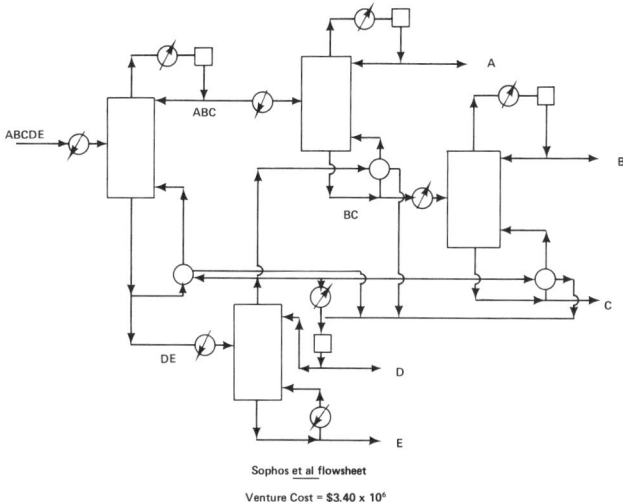

Figure 9. Sophos *et al* flowsheet.

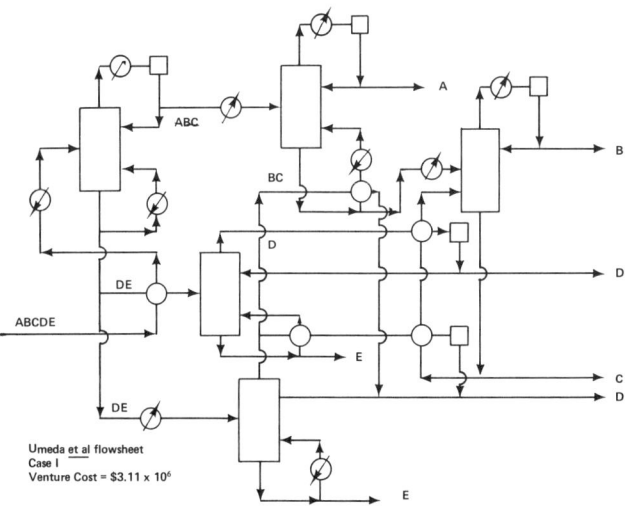

Figure 10. Umeda *et al* flowsheet - Case I.

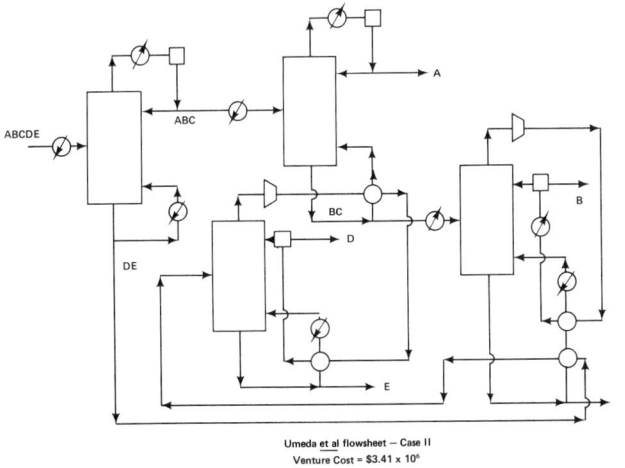

Figure 11. Umeda *et al* flowsheet - Case II.

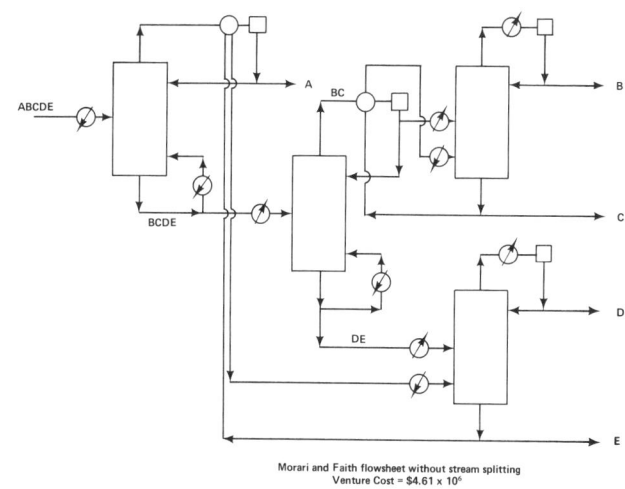

Figure 12. Morari and Faith flowsheet without stream splitting.

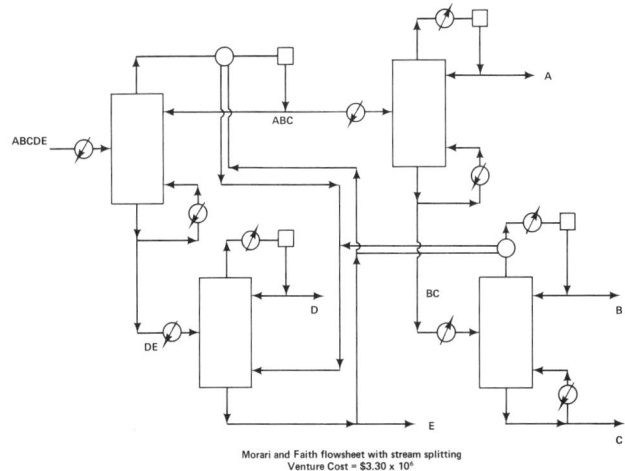

Figure 13. Morari and Faith flowsheet with stream splitting.

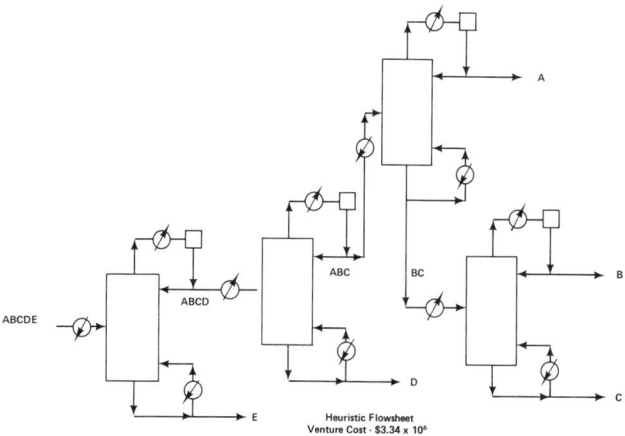

Figure 14. Heuristic flowsheet.

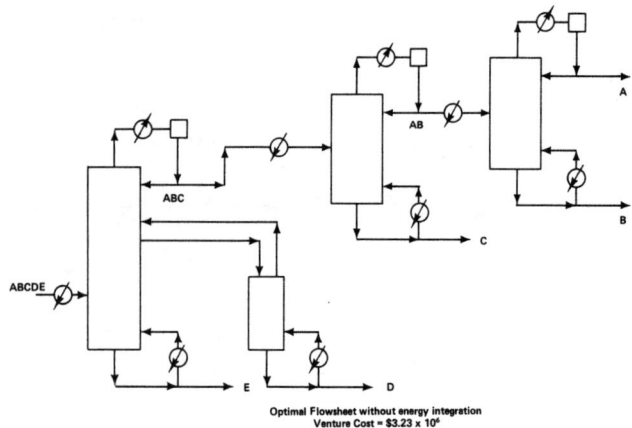

Figure 15. Optimal flowsheet without energy integration.

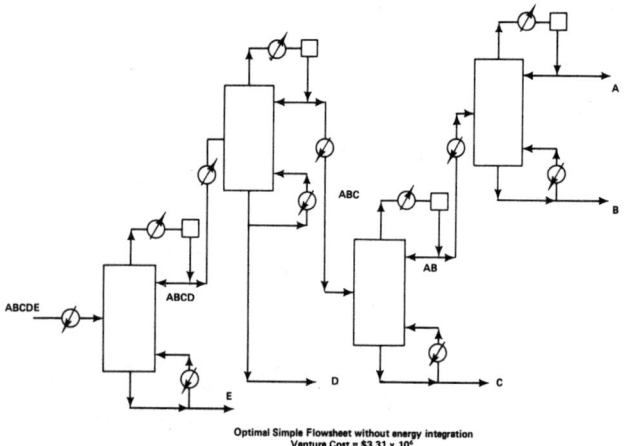

Figure 16. Optimal simple flowsheet without energy integration.

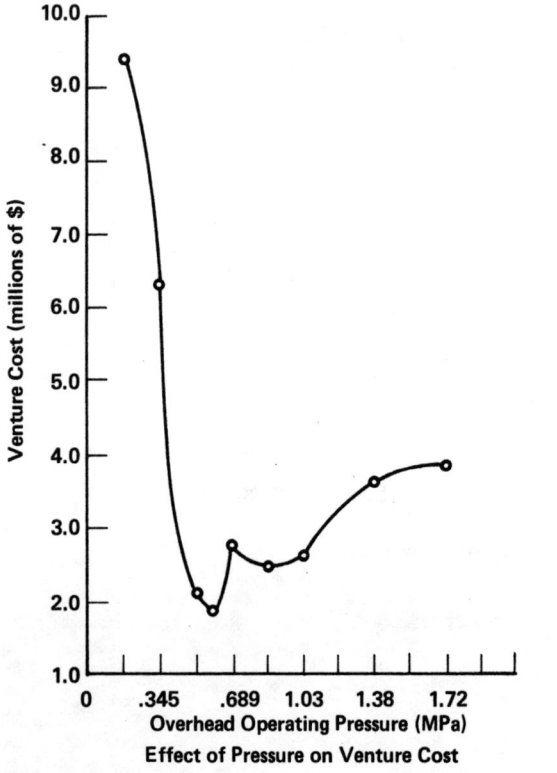

Figure 17. Effect of pressure on venture cost.

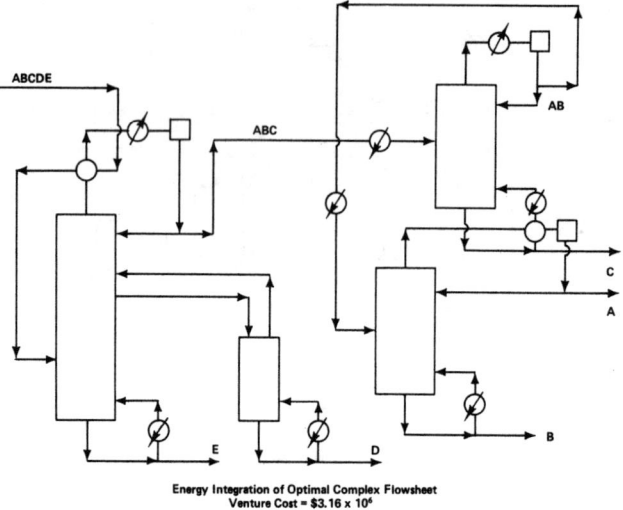

Figure 18. Energy integration of optimal complex flowsheet.

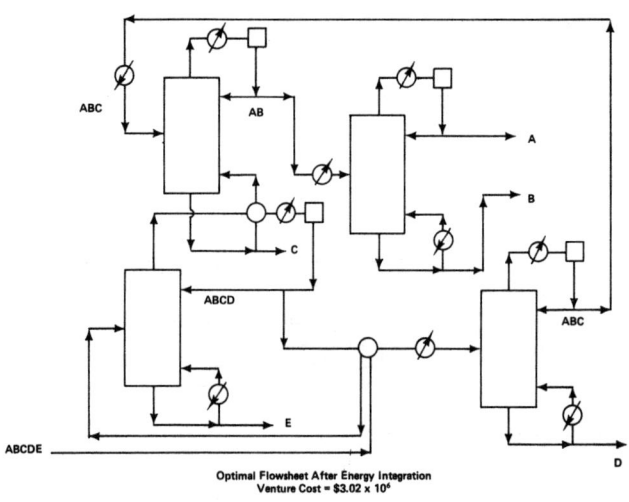

Figure 19. Optimal flowsheet after energy integration.

THE OPTIMAL DESIGN OF RESILIENT HEAT EXCHANGER NETWORKS

A.R. PARKINSON
J.S. LIEBMAN
and
C.O. PEDERSEN

Department of Mechanical and Industrial Engineering
University of Illinois at Urbana-Champaign
Urbana, IL 61801

A.B. TEMPLEMAN

Department of Civil Engineering
The University of Liverpool
Post Office Box 147
Liverpool L69 3BX
Lancaster, England

ABSTRACT

A significant problem in the design of energy efficient thermal systems is the design of resilient heat exchanger networks. A resilient network is able to tolerate fluctuations in stream temperatures and flow rates and still achieve maximum energy recovery while bringing all streams to their target temperatures.

Current design procedures for resilient networks are heuristic in nature and may, therefore, be suboptimal with respect to cost or may not be able to accommodate all stream fluctuations. In the current research, an algorithm is presented which synthesizes networks resilient for all stream fluctuations and minimizes network investment cost for the conditions of maximum energy recovery.

The algorithm relies upon feasibility tests which insure that approach temperatures remain feasible and maximum energy recovery is always achieved. Promising matches are selected on the basis of a lower bound cost estimate of the network. It is postulated that the lower bound can also be used to examine the sensitivity of network cost to limits on stream fluctuations.

A great deal of attention has been devoted in the literature to the design of heat exchanger networks. Traditionally, the design problem has been to find the exchanger network which is optimal with respect to the sum of amortized capital costs (for the exchangers) and the utility (i.e. energy) costs. When the problem was first defined, the combinatorial difficulties associated in solving it were so overwhelming that the following simplifying assumptions were adopted:

1. The cost function for all exchangers was of the form $C = a A^b$ where a and b are constants and A is the area of the exchanger. This equation appears to be based on the work of Guthrie (1969).

2. Only ideal countercurrent exchangers were considered. The area of the exchanger is then given by $A = Q/U\Delta T_L$, where Q is the heat load, U is the overall heat transfer coefficient, and ΔT_L is the log mean temperature difference.

3. Stream properties were assumed to be constant with respect to time and temperature. In this paper, the terms "properties" or "parameters" refer to stream capacity flow rates and inlet and outlet temperatures.

0065-8812-82-6102-0214-$2.00.
© The American Institute of Chemical Engineers, 1982

Considerable progress has been made in developing methodologies which find optimal or nearly optimal solutions to this problem. A review of the work to date may be found in Nishida, et al. (1981). One of the principal findings of the research has been that optimal cost networks achieve a high degree of energy recovery and use a minimum number of exchanger units. Both of these parameters may be determined beforehand. The minimum utility usage may be calculated using the problem table presented by Linnhoff (1978) or the "Heat Availability Function" defined by Raghavan (1977). The minimum number of units is usually equal to the total number of process and utility streams minus one.

Recent research in this area has attempted to broaden the problem definition to make it more realistic. For example, Bharwada and Govind (1981) have included exchanger pressure drop considerations and associated pumping costs in the selection of matches; Challand, et al. (1981), examined temperature difference corrections to account for the use of non-idealized exchangers; the method of Pehler and Liu (1981) can accommodate temperature-dependent properties.

Another important consideration first suggested by Marselle, et al. (1980), and also discussed by Al-Zakri and Bell (1981), is the concept of resilience. Marselle observed that process stream properties in real processes are not perfectly constant but fluctuate with respect to time. A network designed only for one set of stream properties may not only fail to achieve maximum energy recovery but may also fail to bring the streams to their target temperatures much of the time. Obviously, such an inflexible design is of limited value. On the other hand, a network is said to be resilient if it is flexible enough to bring streams to their target temperatures and recover a maximum amount of energy for all stream property variations. The exchangers in such a network are sized for conditions which require maximum exchanger area; a bypass around each exchanger is installed on one of the streams in order to maintain proper stream temperatures when maximum area is not required. The exchanger order, i.e. the network structure, is carefully selected to achieve resilience, and more than the minimum number of exchangers may be necessary.

As might be expected, the design of resilient networks is complex. The combinatorial difficulty of the original problem remains, in addition to which the network must now be able to satisfy many sets of stream conditions.

Marselle, et al., presented a semi-heuristic design procedure which involves synthesizing a network which can handle four extreme stream conditions. Each of the four conditions represents a severe test of network resilience. As they note, however, their method does not guarantee that the network is resilient for all conditions.

In the current research, an algorithm is presented which is capable of generating networks resilient for all conditions. Resilience is guaranteed by feasibility tests which ensure that (1) approach temperatures are positive, (2) target temperatures are met, and (3) maximum energy recovery is achieved for all stream variations. This requirement assumes that the optimal cost network is one which achieves minimum utility usage for all conditions. The minimum utility requirement calculation is embedded directly into the model. Although maximum energy recovery has been a feature of all optimal steady state solutions, the minimization of network cost under the assumption that the optimal utility usage is the minimum possible is only an approximate procedure for arriving at an overall cost optimum. The algorithm selects promising matches for the synthesis on the basis of a lower bound estimate of network cost developed from a mathematical model of a resilient network.

In the next sections, the development of the model, the lower bound, and the feasibility tests are presented. Examples of networks synthesized by the algorithm are given. Finally the limitations of the algorithm and further research being conducted in improving the methods outlined here are discussed. All notation not explained in the text is summarized at the end of the paper.

BACKGROUND

Since this research relies heavily on several major results first developed by Raghavan, his work will be briefly reviewed.

The Heat Availability Function

The Heat Availability Function, or HAF, is of great practical value in assessing the utility needs of a stream set and in determining if a match precludes achieving maximum energy recovery. The HAF is a function of temperature. For any temperature, T, it is defined as the energy available from the hot streams at T minus the energy needed by the cold streams to heat them beyond T to their target temperatures. Or

$$HAF(T) = \sum_i^m \int_T^{T_{max}} C_{hi}(T) dT - \sum_j^n \int_T^{T_{max}} C_{cj}(T) dT \quad (1)$$

An example of the HAF is given in Figure 1. This HAF is for the base stream system taken from Table 1, i.e., the inlet temperatures and capacity flow rates were set at their mean values. The following conclusions may be drawn about the HAF:

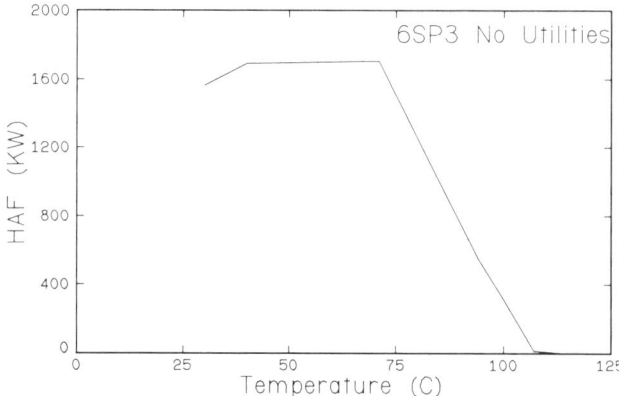

Figure 1 Base HAF for 6SP3. Utility Cooling not Added in.

Table 1 Problem 6SP3 Resilient taken from Marselle, et al. (1980).

Stream		T_{in} (°C)	T_{out} (°C)	C.F. (KW/K)
Hot	1	112-102	40	44.31-36.25
	2	101-89	40	11.12- 9.10
	3	118-108	40	2.25- 1.84
Cold	1	45-35	71	41.86-34.35
	2	35-25	71	13.69-11.33
	3	45-35	99	2.46- 2.02
Cooling Water	4	20	40	131.89-26.19

1. If specific heat is assumed constant with temperature, the HAF is piecewise linear, changing slope at inlet and outlet temperatures of streams.

2. If at any point the HAF is negative, at that temperature more energy is needed by the cold streams than is available from the hot streams. Therefore utility heating is needed at some temperature above this point. The theoretical minimum amount of utility heating is equal to the lowest value of the HAF. The minimum value of the HAF will be raised to zero if only this amount is added to the stream system. This will require a zero temperature of approach in some exchanger. A practical minimum amount of heating is found by assuming a minimum temperature of approach, multiplying this value by the slope of the HAF at its lowest point, and adding this amount (which has the units of energy per unit time) to the theoretical minimum.

When the minimum value of the HAF is negative and occurs at the lowest temperature of the system, the practical minimum utility heating is equal to the theoretical minimum, i.e. only enough heat need be added to raise the HAF to zero. This case will be called "energy deficit heating" to distinguish it from the previous case where heating is required because the hot streams are not at high enough temperature levels to transfer heat to the cold streams. When energy deficit heating occurs, sufficient driving force for heat transfer is available; the hot streams just do not have enough energy to bring the cold streams to target.

3. The amount of utility cooling is equal to the value of the HAF at the lowest temperature of the system (if positive) plus the amount of utility heating, if any.

When the utility heating and cooling have been added into the stream system, the stream set is termed "closed," and the HAF has a value of zero at the minimum and maximum temperature.

The HAF is also important in assessing the impact of a particular match on network energy recovery. If, after a match is made, the HAF of the residual stream set is positive, then maximum energy recovery is still possible because at all temperatures more heat is available than is needed. If, however, after a match the HAF is negative somewhere, then the value of the HAF at its lowest point represents the energy which now must be supplied by extra utility heating. Maximum energy recovery is no longer possible and the match can be rejected.

Developing a Model and a Lower Bound of a Stream System

The HAF of a closed stream set is useful for still another reason. The area under the HAF, F_T, can be used to estimate the cost of the network. For any match between a hot stream i and a cold stream j, an HAF for the match may be constructed. The HAF starts at the highest temperature of the match, the inlet temperature of the hot stream, and ends at the lowest temperature, the inlet temperature of the cold stream. Designating the area under the HAF for the match as F_{ij}, Raghavan discovered that

$$\sum_{i}^{m} \sum_{j}^{n} F_{ij} = F_T, \quad (2)$$

i.e., the sum of the areas of the exchanger HAF's equals the area under the system HAF. Furthermore, Raghavan was able to show the relationship

$$F_{ij} = \Delta T_{Aij} Q_{ij}. \quad (3)$$

If ΔT_{Aij} is used to approximate ΔT_{Lij}, then the area for an exchanger may be written as

$$A_{ij} = Q_{ij}^2 (U_{ij} F_{ij})^{-1}. \quad (4)$$

With this substitution a mathematical model of a stream system may be written as

$$\text{Min } C = a \sum_{i}^{m} \sum_{j}^{n} Q_{ij}^{2b} (F_{ij} U_{ij})^{-b} \quad (5)$$

s.t.
$$\sum_{i}^{m} Q_{ij} = Q_{cj} \quad \forall_j \quad (6)$$

$$\sum_{j}^{n} Q_{ij} = Q_{hi} \quad \forall_i \quad (7)$$

$$\sum_{j}^{n} Q_{cj} = \sum_{i}^{m} Q_{hi} = Q_T \quad (8)$$

$$\sum_{i}^{m} \sum_{j}^{n} F_{ij} = F_T \quad (9)$$

and the Second Law of Thermodynamics. Constraints (6) and (7) state that the heat transferred from/to any stream in several exchangers must equal the total heat available/needed by that stream. Constraint (8) states that the heat transferred across all streams must equal the total heat available, and (9) constrains the sum of the exchanger F_{ij}'s to be equal to F_T. The Second Law enters the problem because the hot streams must be hotter than the cold streams in order to exchange heat.

Assume some distribution of Q's is found which satisfies constraints (6) and (7). The optimal distribution of F's may be found by applying the LaGrange multipler technique to the problem:

$$\text{Min } C = a \sum_{i}^{m} \sum_{j}^{n} Q_{ij}^{2b} (F_{ij} U_{ij})^{-b}$$

s.t. $\sum_i \sum_j F_{ij} = F_T.$ (10)

The solution of the problem, in terms of λ, is

$$\lambda = \frac{b \, U_{max}^{-b} [\sum\sum Q_{ij}^{(2b/1+b)}]^{1+b}}{F_T^{1+b}} \quad (11)$$

and the cost of the network is given by

$$C = \lambda \, ab^{-1} F_T. \quad (12)$$

Thus a lower bound on cost for the network can be calculated if a minimizing set of Q's can be identified. A solution to the model forms a lower bound because the Second Law constraint has been neglected. It should also be pointed out that the arithmetic mean temperature difference used in Eq. (4) is a conservative estimate of the log mean temperature difference; it therefore underestimates the exchanger area needed and makes the lower bound more conservative.

Raghavan obtained a minimizing set of Q's for Eq. (11) by relaxing either constraint (6) or (7) in the model. Obviously a solution which satisfies the relaxed problem forms a lower bound to the actual problem. However, the physical situation represented by the bound is unrealistic and therefore underestimates the optimal cost an average of 20 to 30 percent.

Improving the Lower Bound

A better bound can be achieved by not relaxing constraint (6) or (7). The problem of finding a minimizing set of Q's for Eq. (11) may then be stated as follows:

$$\text{Min} \sum_{i}^{m} \sum_{j}^{n} Q_{ij}^{(2b/1+b)} \quad (13)$$

$$\text{s.t.} \quad \sum_{i}^{m} Q_{ij} = Q_{cj} \quad \forall_j \quad (14)$$

$$\sum_{j}^{n} Q_{ij} = Q_{hi} \quad \forall_i \quad (15)$$

$$\sum_{i}^{m} Q_{hi} = \sum_{j}^{n} Q_{cj} = Q_T \quad (16)$$

The solution to this problem has been determined and implemented into the bound. The optimal set of Q's turns out to be the most uneven distribution possible. A comparison of the new bound and Raghavan's bound is shown in Tables 2 and 3. As can be seen, the new bound is considerably sharper than the old bound, generally averaging within 10 percent of the actual optimal cost. Although the new bound improves algorithm efficiency, its main significance is that it is now sharp enough to be used as an estimate of network cost. It thereby enables the designer to study the sensitivity of network cost to changes in the problem specifications without having to synthesize networks. This could be a powerful tool in estimating trade-offs between energy recovery and capital costs, as for example, in resilient networks.

The bound has also been developed for the case where exchanger heat transfer coefficients or material costs are not constant throughout the network.

DEVELOPING A MODEL AND LOWER BOUND OF A RESILIENT NETWORK

In a resilient network, stream parameters are allowed to fluctuate. Each unique set of stream system parameters with minimum amounts of utility heating and/or cooling added in specifies a closed stream state. Obviously there are an infinite number of

Table 2 Comparison of New Lower Bound to Raghavan's Bound.

	Raghavan's Bound		New Bound		
Prob.[1]	Worst[2] Bound	% Below Opt. Cost	Worst[2] Bound	% Below Opt. Cost	Opt.[3] Cost
4SP1	2163	25	2628	8	2868
4SP2	5358	20	5358	20	6662
4SP3	1976	22	2389	5	2519
6SP1	6005	17	6512	10	7246
6SP2	3236	27	4004	10	4455
7SP1	7065	14	7504	9	8215
10SP1	6949	25	6949	12	9280
6SP3	3814	29	4811	10	5360

[1] For problem definitions, see [9], except for 6SP3 which is defined in Table 1
[2] The worst bound is defined to be the worst bound of any node in the synthesis tree. Thus one match has already been made.
[3] The optimal cost is the annualized exchanger cost only.

Table 3 Comparison of Nodes Evaluated to find Optimal Network

Prob.	New Bound	Raghavan's Bound	Exhaustive Search
4SP1	45	73	106
4SP2	6	6	8
4SP3	10	10	12
6SP1	239	339	545
6SP2	102	114	177
7SP1	242	330	560
10SP1	-	2×10^6	4×10^6
6SP3	520	945	2916

unique stream states. Each stream state requires a certain area for each exchanger in order to satisfy the stream target conditions. Each exchanger must be sized for the stream conditions which require the maximum area. A statement of the objective function for a resilient network may therefore be written as

$$\text{Min} C = a \sum_{ij}^{mn} [\max(A_{ij1}, A_{ij2}, A_{ij3} \cdots A_{ij\infty})]^b \quad (17)$$

where A_{ijk} is the area needed by the match between streams i and j for stream state k. If the substitution of ΔT_A is again made for ΔT_L as was previously done, then a model of a resilient network may be written as

$$\text{Min } C = a \sum_i^m \sum_j^n \text{Max} \{ (\frac{Q_{ij1}^2}{U_{ij1} F_{ij1}})^b \cdots \quad (18a)$$

$$(\frac{Q_{ij2}^2}{U_{ij2} F_{ij2}})^b \, (\frac{Q_{ij\infty}^2}{U_{ij\infty} F_{ij\infty}})^b \} \quad (18b)$$

s.t.

$$\sum_j^n Q_{ijk} = Q_{hik} \quad \forall i,k \quad (19)$$

$$\sum_i^m Q_{ijk} = Q_{cjk} \quad \forall j,k \quad (20)$$

$$\sum_i Q_{hik} = \sum_j Q_{cjk} = Q_{Tk} \quad \forall k \quad (21)$$

$$\sum_i^m \sum_j^n F_{ijk} = F_{Tk} \quad \forall k \quad (22)$$

and the Second Law of Thermodynamics. The cost exponent "b" can be pulled inside the braces because it is greater than zero.

Instead of trying to evaluate all possible stream conditions, suppose that a finite number of stream states are chosen which represent a good cross section of possible states. If a minimizing set of Q_{ijk}'s for these states can be found which satisfy (19) and (20), what is the optimal distribution of F_{ijk}'s?

For purposes of discussion, assume that only two states are involved. For each state, distribute the F_{ijk}'s according to the optimal distribution for a single stream state system. This creates two estimates of the area required for each lower bound match—one for each stream state. Assume for the first match that the area needed by state 1 is greater than the area needed by state 2, or

$$Q_{11}^2/F_{11} > Q_{12}^2/F_{12} \quad (23)$$

where the first subscript now refers to the match number and the second subscript refers to the stream state. Also all heat transfer coefficients are assumed equal and therefore do not appear in the area estimates. The estimated area needed for the match is

$$\text{Max}\{(Q_{11}^2/F_{11}), (Q_{12}^2/F_{12})\} = Q_{11}^2/F_{11}. \quad (24)$$

Notice that F_{12} may be decreased at no cost until either $(Q_{12}/F_{12}) = (Q_{11}/F_{11})$ or $F_{12} = Q_{12}$, which is the smallest F_{12} can become. (Practically it can never become this small since this would require an approach temperature of one degree at both ends of the exchanger.) By decreasing F_{12}, F_{p2}'s (where p is the match number) of other matches may be increased, decreasing those exchanger areas. To specify this procedure in general terms: for each match pair, designate the set composed of all areas which are smallest of the pair and belong to the same stream state the set "S_k." Call the set containing all areas which are the largest in the pair the set "L_k." Decrease the F's in each set S_k to their lower limits M_{pk}. Subtract the sum of these constants from F_{Tk}, i.e.

$$F'_{Tk} = F_{Tk} - \sum_{p \in S_k} M_{pk}. \quad (25)$$

The problem for obtaining the optimal distribution of F_{ijk}'s can then be written

$$\text{Min } z = (a/U^b)[\sum_{p \in L_1} \frac{Q_{p1}^2}{F_{p1}}^b + \sum_{p \in L_2} \frac{Q_{p2}^2}{F_{p2}}^b] \quad (26)$$

$$\text{s.t.} \quad \sum_{p \in L_1} F_{p1} = F'_{T1} \quad \sum_{p \in L_2} F_{p2} = F'_{T2}. \quad (27)$$

Since this problem is separable, the optimal distribution of F's is the optimal distribution for each system. Thus a lower bound for the system is

$$C = a \, U^{-b} \left(\{ [\sum_{p \in L_1} (Q_{p1}^{(2b/1+b)}/F'_{T1}]^{1+b} + [\sum_{p \in L_2} (Q^{(2b/1+b)}/F'_{T2}]^{1+b} \right\} \quad (28)$$

The problem remains of choosing a minimizing set of Q's. This problem is much more complex than the selection of Q's for the single state case and no solution has been found. Rather an heuristic has been used to select Q's. The heuristic is to select a heat load according to the criterion (and assuming only two stream states are used for the bound):

$$\text{Min } \frac{|Q_{iji} - Q_{ij2}|}{\text{Max } \{Q_{iji}, Q_{ij2}\}} \quad (29)$$

This heuristic is based on the concept that if an exchanger must be sized for a certain heat load for one stream state, it would be highly desirable to carry the same load in other states in order to minimize the sizes of exchangers derived from loads of other states. Since the heuristic may not find the optimal set of Q's, the lower bound estimate might be higher and sharper than it actually should be. If the heuristic is poor, the lower bound could exceed the actual optimal cost. However, this heuristic appears to be justified not only intuitively but by experimental results as well. In all cases examined so far, the bound has been observed to gradually sharpen as more matches are made, and in no case has the bound ever exceeded the actual optimal cost. Preliminary results indicate the bound to average within 15 percent of the actual optimal cost.

The algorithm uses three states in the generation of the bound: (1) energy available from the hot streams set at a maximum, energy needed by the cold streams set at a minimum (maximum cooling needed), (2) energy available from the hot streams set at a minimum, energy needed by the cold stream set at a maximum (maximum heating needed) and (3) streams set to test for energy feasibility--a state which will be described in the following section. The first two states are equivalent to two of the corner points used by Marselle.

It should be emphasized that these three states are used only to generate an estimate of network cost and not to determine the resilience of specific matches. The resilience of specific matches is determined by the feasibility tests described in the next section. The tests are designed to check resilience for the entire range of match variable fluctuations and not just for three states.

FEASIBILITY TESTS

Energy Recovery Test

In the introduction, the use of the HAF in checking for energy recovery for a single stream state was described. If, after a match, the residual stream HAF is negative, then that match precludes maximum energy recovery and is rejected. It should be mentioned that the residual stream HAF will differ from the HAF existing before the match only in the match temperature interval; hence only the match interval need be checked for negative values. The single state test has a natural extension to resilient networks: if, after a match, the residual HAF is negative in the match interval for any possible stream state, then that match prevents complete energy recovery for that state and can be rejected. Since it is computationally infeasible to examine all stream states, a substitute criterion, the worst average HAF, or WAHAF, is used. The WAHAF is defined to be the HAF which has the lowest average value in the match interval. A methodical procedure for specifying the stream state which produces the WAHAF has been developed and is discussed next.

Given a stream system in some arbitary stream state, the state and therefore HAF may be changed by adjusting stream parameters within allowable limits. At all times, however, the stream system must be in energy balance. Thus a change in a hot stream which increases its energy content, for example, must be balanced by either a commensurate increase in a cold stream or a decrease in another hot stream. This points out the two kinds of changes which can be made: adjustments can be made among streams of the same type (called hereafter a stream set), in which case energy is merely reallocated within the set; or adjustments can be made across sets in which case the energy content of both the hot and cold sets is changed by the same amount.

An effective strategy for making adjustments is facilitated by the theorems below. Details of the proofs are available but are not included in the paper.

1. Theorem 1. The same final stream state can be arrived at by making stream adjustments in any order.

2. Theorem 2. Suppose both the cold and hot stream sets have been adjusted so that no further reduction in the HAF is possible by making adjustment within sets. Then if an adjustment match is made between sets which effects the largest reduction in the HAF possible, no further reduction within sets is possible.

3. Theorem 3. Adjusting stream levels is exactly analogous to making a heat exchanger match between the portions of the streams which are adjusted. Accordingly, the net average change in the HAF caused by an adjustment match (denoted by the subscript "am") is given by

$$\frac{Q_{am} \Delta T_{Aam}}{T_{am_{max}} - T_{am_{min}}} \qquad (30)$$

Using these theorems, the WAHAF may be found by the following procedure:

1. Set all streams to their lowest energy values. Place the hot and cold stream sets in energy balance by raising the energy content of the smallest set. This increase is made by changing those stream parameters which effect the largest negative change (for the cold streams) or smallest positive change (for the hot streams) in the HAF as defined by Theorem 3. No further adjustments are possible which lower the HAF within sets.

2. Make adjustments between sets in the order specified by Theorem 2. When no further reduction is possible, the WAHAF has been found.

The WAHAF for the stream system of Table 1 is presented in Figure 2. The dotted line is the HAF for the base case. Because the WAHAF is everywhere positive, initially, at least, maximum energy recovery is possible for all states.

The WAHAF in Fig. 2 is identical to the "maximum heating case" defined by Marselle. No utility heating is indicated by the WAHAF, contrary to the results of Marselle, because no minimum temperature of approach was assumed. (Marselle assumed a minimum temperature of approach of (10°C). Generally, however, the WAHAF is not equivalent to the "maximum heating case" of Marselle. The WAHAF is the state that on the average provides the most severe test of energy recovery. This state is problem dependent and cannot be easily predicted beforehand, but is derivable using the above theorems. The WAHAF for the stream system of 5SP1, given in Table 4, for example, is a state which requires neither heating nor cooling, i.e. the streams are in perfect energy balance.

TEMPERATURE FEASIBILITY TEST

The temperature feasibility test consists of verifying that the approach temperatures for a match are positive for all fluctuations. A discussion of the test is facilitated by first explaining the types of matches synthesized by the algorithm.

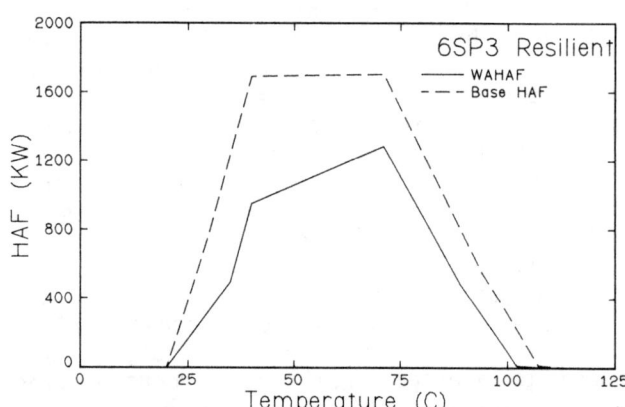

Figure 2 WAHAF and Base HAF for 6SP3. Utility Cooling Added in.

Table 4 Problem 5SP1 Resilient

Stream		T_{in} (°C)	T_{out} (°C)	C.F. (KW/K)
Hot	1	253.9-243.9	121.1	18.28-14.96
	2	209.4-199.4	65.6	14.62-11.96
Cold	1	42.8- 32.8	204.4	12.53-10.25
	2	70.6- 60.6	182.2	14.35-11.63
	3	98.3- 88.2	204.4	14.33-11.73
Cooling Water	4	37	80	7.71- 0

The algorithm is constrained to look for matches in which one stream is brought to its target temperature. The three types of such matches considered by the algorithm are given in Fig. 3. The ordinates in the figure are in relative units so that the area enclosed by the curves represents the energy available or needed by a stream. Only one state is shown for the streams. It can be imagined that the area expands and contracts as fluctuations occur and the energy content of the streams change. It is arbitrarily assumed in the figure that the cold stream is brought to target. The hottest part of the hot stream is matched in Fig. 3b. This match has largest possible mean temperature difference. An intermediate temperature difference is produced by stream splitting, shown in Fig. 3c. The least possible temperature difference is achieved when the coldest part of the hot stream is matched to the cold stream, as in Fig. 3d.

The feasibility test involves checking the approach temperatures at both ends of the exchanger. Suppose, for example, the type of match shown in Fig. 3b is proposed between a hot stream k and a cold stream g. It will be assumed that the hot stream brings the cold stream to target for all fluctuations. (In general, however, this assumption need not be made. The hot stream might alternate between bringing the cold stream to target and being brought to target. Such a match leaves two residual streams and leads to networks with more than the minimum number of units). The minimum approach temperature at the high temperature end of the exchanger is easily found by inspecting the limits on T_{hki} and T_{cog}. Writing the upper and lower limits on these variables as $\overline{T}$ and $\underline{T}$, respectively, if $\underline{T}_{hki} > \overline{T}_{cgo}$, then this approach temper-

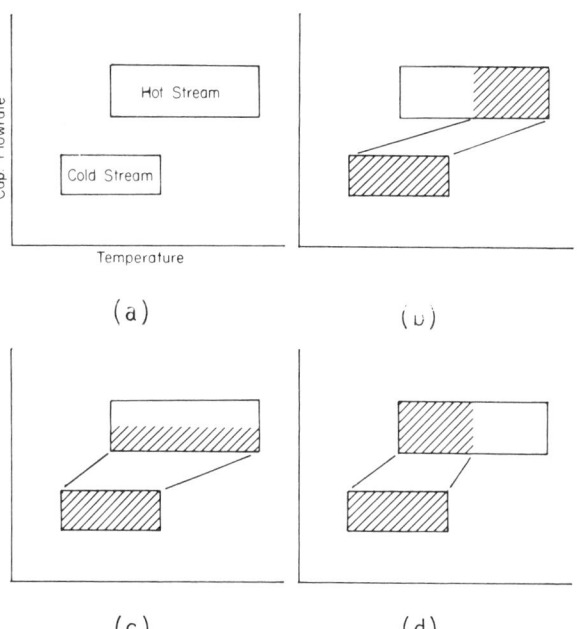

Figure 3 Types of Matches Considered by Algorithm. (a) Streams Matched; (b) Hottest Part of Hot Stream Matched to Cold Stream; (c) Hot Stream Split; (d) Coldest Part of Hot Stream Matched to Cold Stream (after Raghavan, 1977).

ature is always positive. The approach temperature at the cold end of the exchanger is more difficult to solve for because T_{hko} for this kind of match is given by

$$T_{hko} = T_{hki} - C_{cg}/C_{hi}(T_{cgo} - T_{cgi}) \quad (31)$$

Thus, the approach temperature test takes the form:

Min $(T_{hko} - T_{cgi})$, s.t. Eq. (31)

and upper and lower limits on all variables.

Unfortunately, the problem is complicated by the fact that in some cases, notably when a utility stream is involved in a match, the variables are not free to fluctuate independently of each other. This dependence can be understood by considering the energy balance the hot and cold streams must maintain. Suppose, as in problem 6SP3, only utility cooling is necessary, and cold stream g is the utility stream. Then the minimum utility requirement is given by:

$$\sum_{i}^{m} Q_{hi} - \sum_{j \neq g}^{n} Q_{cj} = Q_{cg}^{*} \quad (32)$$

where the asterisk indicates stream g is the utility stream. The upper and lower limits on the utility stream are given by

$$\sum_i \overline{Q}_{hi} - \sum_{j \neq g} \underline{Q}_{cj} = \overline{Q}^*_{cg} \qquad (33)$$

$$\sum_i \underline{Q}_{hi} - \sum_{j \neq g} \overline{Q}_{cj} = \underline{Q}^*_{cg} \qquad (34)$$

It can be seen that maximum cooling occurs when the hot streams are high and the cold streams are low, and vice versa for minimum cooling. Continuing the example, the residual stream left after hot k contacts cold utility g in a match is given by:

$$Q_{res} = Q_{hk} - Q^*_{cg} \qquad (35)$$

If these streams could fluctuate independently, the limits on Q_{res} would be:

$$\overline{Q}_{res} = \overline{Q}_{hk} - \underline{Q}^*_{cg} \qquad (36)$$

$$\underline{Q}_{res} = \underline{Q}_{hk} - \overline{Q}^*_{cg} \qquad (37)$$

However, these limits violate Eqs. (33) and (34). This can be seen by substituting $(Q_{hk} + \sum_{i \neq k} Q_{hi}) = \sum_i Q_{hi}$ into Eqs. (33) and (34) and rearranging to give

$$(\sum_{j \neq g} \underline{Q}_{cj} - \sum_{i \neq k} \overline{Q}_{hi}) = \overline{Q}_{hk} - \overline{Q}^*_{cg} \qquad (38)$$

and

$$(\sum_{j \neq g} \overline{Q}_{cj} - \sum_{i \neq k} \underline{Q}_{hi}) = \underline{Q}_{hk} - \underline{Q}^*_{cg} \qquad (39)$$

The left term in Eq. (38) is the smallest this term can become. The left term in Eq. (39) the largest this term can become. The right term of both equations is the expression for Q_{res}. Therefore,

$$\overline{Q}_{res} = \overline{Q}_{hk} - \overline{Q}^*_{cg} \qquad (40)$$

$$\underline{Q}_{res} = \underline{Q}_{hk} - \underline{Q}^*_{cg} \qquad (41)$$

By comparing Eqs. (36) with (40) and (37) with (41), it is obvious that the requirement of maintaining an energy balance constrains $\overline{Q}_{res}$ and $\underline{Q}_{res}$ to be smaller (larger) than they would normally be. Thus the original problem of finding the minimum approach temperature may be written as

Min $(T_{hko} - T_{cgi})$

s.t. $T_{hko} = T_{hki} - C_{cg}/C_{hi}(T_{cgo} - T_{cgi})$

$\underline{Q}_{res} \leq Q_{hk} - Q_{cg} \leq \overline{Q}_{res}$

and upper and lower limits on all variables.

It turns out that the above minimization problem has an underlying structure which can be exploited to solve it quickly using derivative information.

An interesting feature of utility matches can be seen by combining Eqs. (38), (41), (39) and (40) to give

$$\sum_{j \neq g} \underline{Q}_{cj} - \sum_{i \neq k} \overline{Q}_{hi} = \underline{Q}_{res} \qquad (42)$$

$$\sum_{j \neq g} \overline{Q}_{cj} - \sum_{i \neq k} \underline{Q}_{hi} = \overline{Q}_{res} \qquad (43)$$

It can be seen that the residual hot stream now acts just as a utility stream in keeping the streams in energy balance--it fluctuates high when the cold streams are high and hot streams low, and vice versa. If a cold stream now brings the hot residual to target, the cold residual becomes a "utility" stream, and in this manner "utility" streams can be propagated through the network. Although the utility match constraint complicates the determination of dependent match variables, the constraint has the beneficial effect of moderating the limits of the variables to be less extreme than in a regular match. This makes it possible for more matches to be feasible.

ALGORITHM DESCRIPTION

The synthesis algorithm strategy is commonly referred to as depth first branch and bound. The algorithm starts by generating all possible matches which pass the feasibility tests. The matches are rank ordered according to the lower bound and stored in a stack, with the match having the least lower bound on top of the stack. This match is chosen for further expansion. The stream set is updated to reflect the changes in the streams caused by the match. All possible matches are again generated, rank ordered, and stacked. The least lower bound at any stage is the sum of the cost of the synthesized portion of the network plus the lower bound on cost of the residual stream set. The cycle repeats itself until either an entire network is synthesized or no more feasible matches can be found (called "fathoming"). In both cases, the tree is then backtracked by pulling matches off the stack. If a network has been created and the network cost is below the lower bound of the match pulled off the stack, the match is "pruned," i.e. discarded, because it is known that expansion of that match cannot lead to a less expensive network. In this fashion only promising matches are expanded. The synthesis ends when all matches have either been pruned or expanded. A flow diagram summarizing the different steps of the algorithm is given in Fig. 4.

DISCUSSION OF EXAMPLES

Three examples of networks synthesized by the algorithm are given in Figs. 5, 6, and 7. Inlet temperature fluctuations of +5°C and capacity flow rate variations of +10 percent were assumed for both problems. The design data for the problems are given in Table 5. Stream target temperatures were considered constant. The synthesis tree was not exhaustively searched for either of the networks. However, the first network is at least within 2 percent of the optimal cost; the second network is within 5 percent.

A solution for problem 6SP3 is presented in Fig. 5. This can be compared to the solution presented by Marselle. Although he did not estimate cost, it can be seen that the current solution uses two less exchangers. One of the exchangers eliminated is a steam heater; a heater is not needed if a minimum approach temperature, or MAT, of 3°C is allowed. As previously mentioned, Marselle always assumed an MAT of 10°C. An MAT is required by the algorithm only when utility heating and cooling occur simultaneously, as in 4SP1.

The annualized cost of the exchangers (utility costs not added in), $7,361, is 37 percent above the baseline cost of $5,360. Interestingly enough, a solution requiring only six exchangers, shown in Fig. 6, is slightly more expensive at $7,569.

A resilient network for problem 5SP1 is given in Fig. 7. The data for the problem are given in Table 4. This system alternately requires both heating and cooling and requires at least six exchangers. The addition of resiliency to the baseline network increased the exchanger cost from $5,428 to $6,776.

The networks of both problems were simulated in order to verify their resiliency and controllability. All variables were assumed to have a uniform probability distribution between upper and lower limits. By employing a uniform distribution, extreme points are just as likely to occur as midrange points. Nevertheless, the probability of all conditions which require maximum exchanger area existing simultaneously is so small that in 5000 simulations the maximum area was not needed in any one exchanger. This indicates that exchanger areas should be sized according to confidence limits placed on the area distribution functions.

ALGORITHM LIMITATIONS AND FURTHER RESEARCH

The modest fluctuations of +5°C on inlet temperatures and +10 percent of flow rates can introduce large swings in residual stream parameters and make it difficult for feasible matches to be generated. For some problems, e.g. 6SP1, these fluctuations are severe enough that a feasible network cannot be found by the algorithm. Further research will investigate other synthesis strategies which are more robust in situations where very few feasible matches exist.

The applicability of the algorithm would also be enhanced if temperature dependent properties could be considered. All of its underlying principles--the HAF, lower bound, and feasiblity tests, are valid for temperature dependent capacity flow rates. The only difficulty in relaxing this assumption is in developing efficient routines to handle the added computational complexity.

The networks presented in Figs, 5, 6, and 7 are optimal only in the sense that they represent the lowest cost (or nearly so) networks which achieve maximum energy recovery for all conditions. Network simulation has already shown that some conditions occur so infrequently that the exchangers could be downsized, in some cases as much as 20 percent, with very little loss in resiliency. The solution methodology will be extended to include sizing the exchangers according to confidence limits placed on the area distribution functions. The lower bound cost estimate of a network will also be used to examine network cost sensitivity to changes in the limits of stream fluctuations.

Finally, stream target temperatures will be allowed to float in order to further reduce utility consumption or minimize exchanger area.

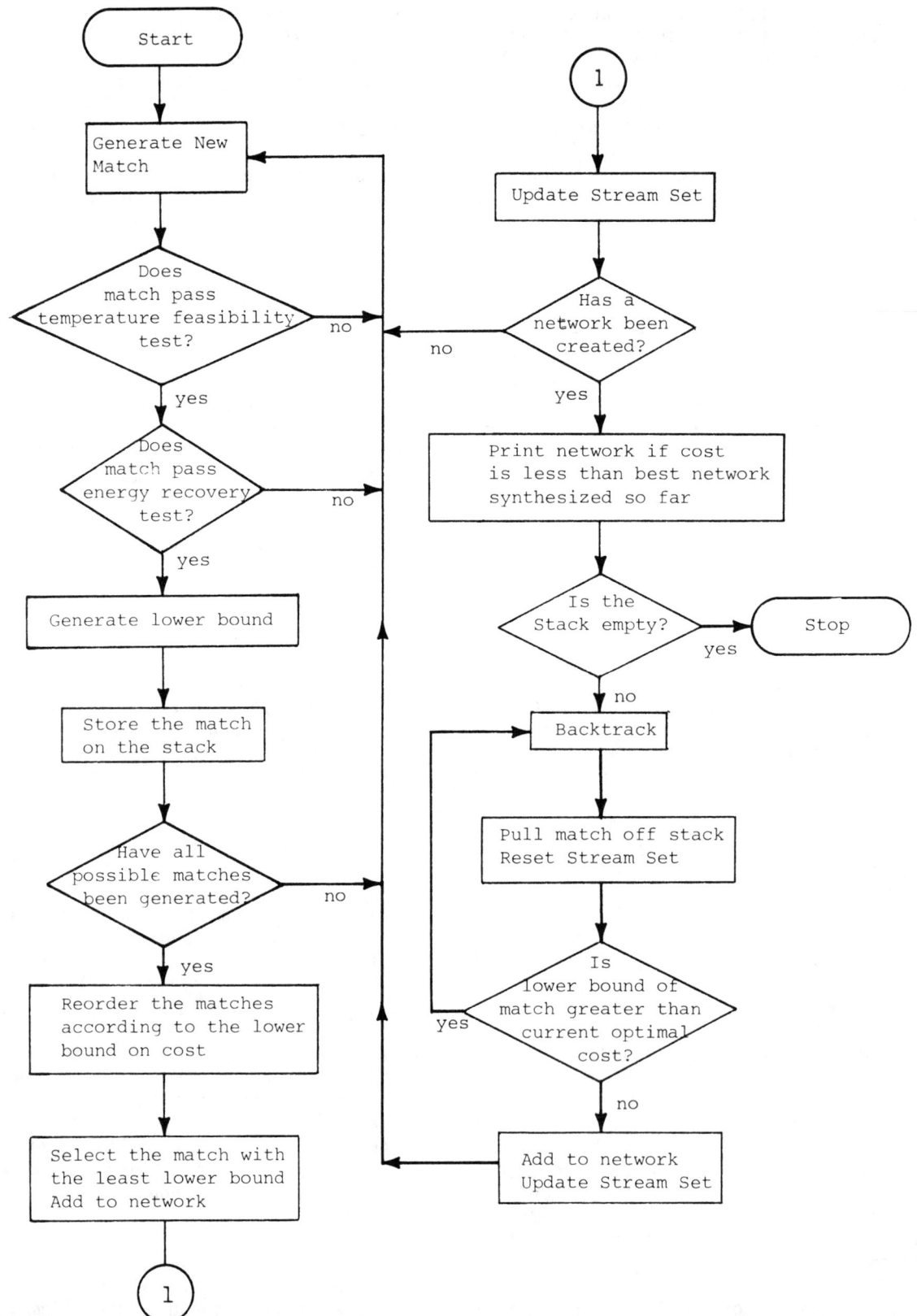

Figure 4 Flow Diagram of Algorithm Operation

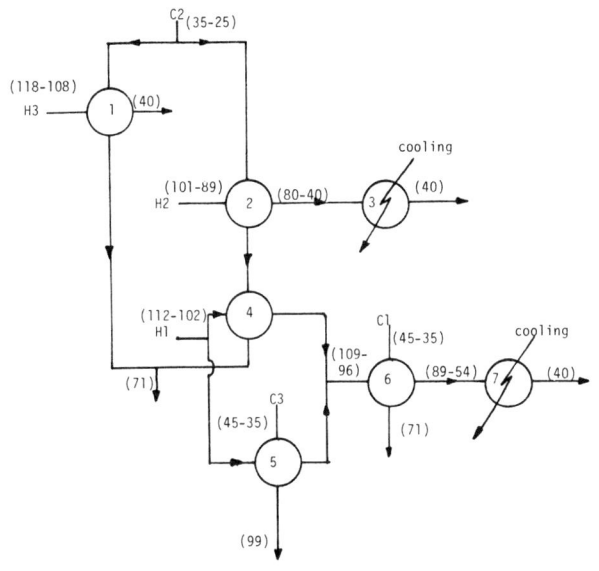

Figure 5 Optimal Resilient Network for 6SP3. Annualized Cost of Exchangers: $7,361.

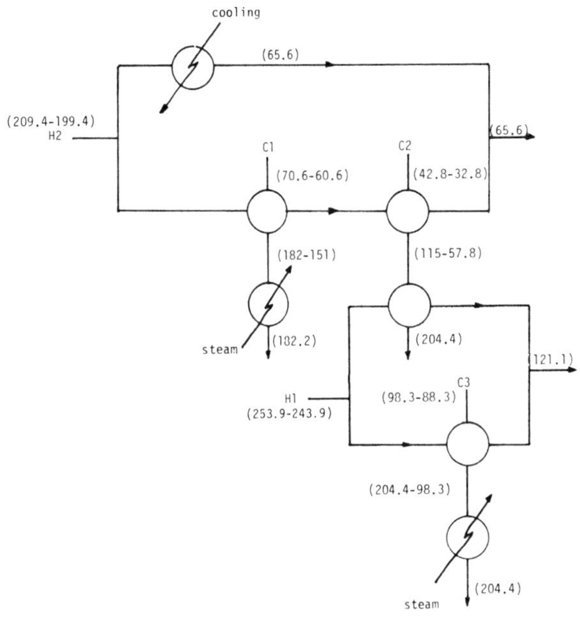

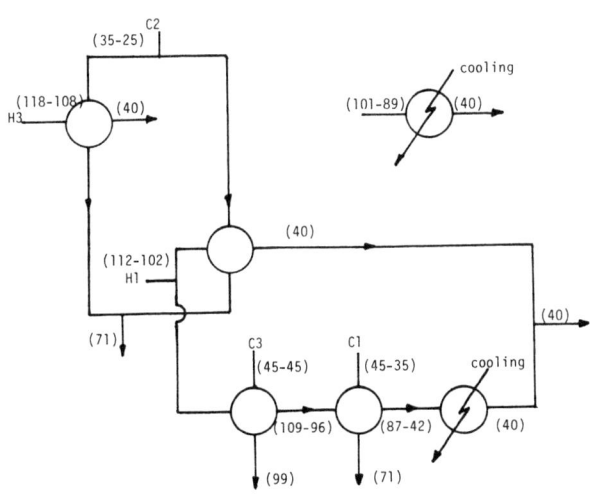

Figure 6 Resilient Network for 6SP3 Using only Six Exchangers. Cost: $7,569.

Figure 7 Resilient Network for 5SP1.

Table 5 Design Data

Overall heat transfer coefficients

Steam heater - 1135.61 W/m^2/K
All others   - 851.74 W/m^2/K

Coefficients for cost equation:

a = 1,456.3
b =   0.6

Annual rate of return = 0.1

Temperature of Steam = 235.5 °C

NOTATION

a	cost coefficient
A_{ij}	area of exchanger (hot i matches cold j)
U_{ij}	overall heat transfer coefficient (hot i matches cold j)
b	cost exponent
C_{hi}	capacity flow rate hot stream i
C_{cj}	capacity flow rate cold stream j
F_{ij}	area under HAF for exchanger (hot i matches cold j)
F_T	total area under HAF for stream system
m	number of hot streams including hot utilities
n	number of cold streams including cold utilities
Q^*	energy of utility steam
Q_{cj}	energy needed by cold stream j
Q_{hi}	energy available from hot stream i
Q_{res}	energy of residual stream
Q_T	total energy available from hot streams and needed by cold streams (with minimum utility requirements included)
ΔT_{Lij}	log mean temperature difference
ΔT_{Aij}	arithmetic mean temperature difference
T_{hii}	inlet temperature of hot stream i
T_{hio}	outlet temperature of hot stream i
T_{hit}	target temperature of hot stream i
T_{cji}	inlet temperature of cold stream j
T_{cjo}	outlet temperature of cold stream j

LITERATURE CITED

1. Al-Zakri, A. S., and K. J. Bell, "Estimation of the Performance of a System of Heat Exchangers when Uncertainties Exist," presented at AIChE Spring Nat. Meeting, Houston, Texas (April 1981).

2. Bharwada, J. J., and R. Govind, "A Non-Iterative Approach to the Synthesis of Optimal Heat Exchanger Networks," presented at AIChE Spring Nat. Meeting, Houston, Texas (April 1981).

3. Challand, T., R. W. Colvert, and C. K. Venkatesh, "Computerized Synthesis Optimization of Process Heat Exchanger Networks," presented at AIChE Spring Nat. Meeting, Houston, Texas (April 1981).

4. Guthrie, K. M., Chem. Eng., p. 144 (24 March 1969).

5. Linnhoff, B., and J. R. Flower, AIChE J., 24(4), 633 (1978).

6. Marselle, D. F., M. Morari, and D. F. Rudd, "Design of Resilient Processing Plants II, Design and Control of Energy Management Systems," presented at AIChE Annual Meeting, Chicago, Ill. (1980).

7. Nishida, N., G. Stephanopoulos, and A. W. Westerberg, AIChE J., 27(3), 334 (1981).

8. Pehler, F. and Y. A. Liu, "Studies in Chemical Process Design and Synthesis: A Simple Thermoeconomic Approach to the Multiobjective Synthesis of Heat Exchanger Networks," presented at AIChE Summer Nat. Meeting, Detroit, Mich. (Aug. 1981).

9. Raghavan, S., "Heat Exchanger Network Synthesis: A Thermodynamic Approach," Ph.D. Dissertation, Purdue University (1977).

PROCESS OPTIMIZATION WITH THE ADAPTIVE RANDOMLY DIRECTED SEARCH

D.L. MARTIN
J.L. GADDY

Department of Chemical Engineering
University of Arkansas
Fayetteville, AR 72701

ABSTRACT

Random search procedures have been applied to the optimization of a number of process problems. These procedures, although somewhat inefficient, have been demonstrated to be highly successful in solving these large, heavily constrained problems. A new algorithm, called the Adaptive Randomly Directed Search, has been developed to improve the efficiency of these procedures by utilizing the history of the search to guide the step size. This technique randomly samples from the surface of an n-dimensional ellipsoid. The size of the ellipsoid, and hence the step size, is controlled by prior successes of the search. This paper describes the function of this new algorithm and presents solutions to several example problems. The efficiency is compared with other methods from the literature used to solve these same problems.

INTRODUCTION

The optimization of a large chemical process is a difficult problem of high dimensionality, often with both real and integer independent variables, and with numerous constraints, many of which are implicit. Due to the complexities of such problems, most studies have used simplified cases.

Parts of processes, such as reactors (Barneson, et al., 1970), absorber-strippers (Umeda and Ichikawa, 1971) and distillation column-condenser systems (Zellnik, et al, 1962) have been studied using sophisticated optimization algorithms. However, optimization of the individual units of a process usually does not yield the same result as optimization of the combination of units in the process. Simplified processes have also been studied (Braken and McCormick, 1968; Gottfried, et al, 1970).

Perhaps the most common technique used is to describe the system and constraints with a set of equations which can be linearized, and to solve the set of linearized equations simultaneously using linear programming. In the same fashion, the set of equations may be put in the form necessary to use any of the generalized programming techniques, such as geometric programming. Recently, a new technique (Westerberg, et al, 1980) solves the large problem with an algorithm which is a combination of a Newton-Raphson method and quadratic programming. This technique linearizes the system equations and constraints and then solves the set simultaneously. In the technique, Lagrangian functions are used to handle the constraints. The accuracy of these solutions depends upon the accuracy of the system of equations used to linearize the process.

Several studies have been made on the optimization of large complex processes using flow sheet simulators. PACER was used by Shannon, et al, (1966) to study the sulfuric acid process and by Seader and Dallin (1973) to study the toluene dealkalation process. Friedman and Pinder (1972) optimized a gasoline polymerization unit, modeled with CHESS. PROPS has been used to study a number of processes: gasoline polymerization, (Gaines and Gaddy, 1976); methanol, (Ballman and Gaddy, 1976); ethylene, (Jinkerson and Gaddy, 1979); and sulfuric acid, (Doering and Gaddy, 1980).

The studies using PROPS have utilized the Adaptive Random Search (ARS) algorithm. This procedure, suggested by Gall (1966), has been found to be quite effective for application to these large heavily constrained problems. The basic algorithm is described by:

$$x_i = x_i^* + R_i (2\theta_i - 1)^k$$

where:
- x_i = new value of independent variable i ($i=1,2,...,n$; n = number of independent variables)
- x_i^* = value of variable x_i which has produced the highest objective function y^*
- R_i = allowable search region (range) about x_i^*
- θ_i = random number between zero and one
- k = distribution coefficient (odd integer 1,3,5,...)

The search randomly samples from the symmetrical distribution, $(2\theta - 1)^k$, for new and better search points. The search region is centered about the best values of the independent variables and the region moves toward the optimum as improved values of the variables are located. The efficiency of the search is controlled by increasing the distribution coefficient, k, as the optimum is approached. Higher values of k reduce the search region and increase the probability of choosing successful new search points, thereby decreasing the step size. This technique has been proven effective in following constraints and has demonstrated superior efficiency in comparison with other algorithms.

The ARS algorithm was modified by Heuckroth, et al, (1976) to reduce the range by dividing by the distribution coefficient. Systematic range reduction greatly improved the efficiency of the technique by decreasing the search step size more rapidly as the optimum was approached. Skewing the distribution to keep larger portions of the distribution in the feasible region when near a constraint was also somewhat helpful. The ARS can be readily adapted to discrete and mixed integer optimization problems, (Kelahan and Gaddy, 1978). Bounds testing and constraint relaxation in the early stages of the search were found to be beneficial in heavily constrained problems (Jinkerson and Gaddy, 1979).

These studies all served to demonstrate the capabilities of ARS and to improve its efficiency. Further improvements are considered possible by using different coefficients to alter the distribution and range of each independent variable. Although convergence is usually guaranteed, the efficiency of the algorithm is dependent upon the experience of the user. The choice of the initial range and the frequency of incrementing the coefficients is specified by the user. Furthermore, a more direct means of controlling the step size would likely improve the efficiency.

The objectives of this study were to examine the ARS to determine the modifications that would improve the efficiency and make the algorithm user independent. A means of using the search history to guide the step size and the selection of the range and distribution coefficients was developed. This new algorithm, Adaptive Randomly Directed Search, was applied to several examples from the literature to determine its utility and efficiency. The algorithm was also applied to a large design problem to insure that the effectiveness in solving more complicated problems was retained or improved.

RANDOM SEARCH PROCEDURES

Pure random search was first proposed by Brooks (1958). The objective function was measured at N random points selected from a uniform probability distribution over the entire parameter space and the best value of the objective function was taken as the optimum. Sprang (1962) showed that in order to achieve a probability of 0.9 of locating the optimum, the number of trials, N, would have to be approximately 2.3×10^n (where n is the dimensionality) for cases where $10^n \gg N$. This large number of trials prohibited the practical use of pure random search, even though finding the optimum was assured if enough trials are made.

Many other pseudo random techniques (Matyas, 1965; Price, 1977; Schumer, 1968; Lawrence, 1973; Wozny and Heydt, 1972; Luus and Jaakola, 1973; and Luus and Wang, 1978) have been developed to improve the efficiency of pure random search. In general, these methods modified the pure random search by altering the allowable search region. Lawrence (1973) presented a randomized pattern search and Price (1977) developed a controlled random search. Example problems were solved by Luus and Jaakola (1973) and Luus and Wang (1978) which demonstrated significant improvements in efficiency over pure random search. Accelerating pure random search increases the risk of convergence to a local optima.

The use of the search history to accelerate the convergence has been studied in a few cases. The mose successful method developed is positive and negative biasing (Bekey, et al, 1966). This simple modification repeated successful trials and used the reverse of unsuccessful trials. Lawrence (1973) showed that this procedure resulted in a significant improvement in the search efficiency. Other, more complex, methods of biasing, such as the use of a transformation matrix (Matyas, 1965) and theoretical optimal step size (Schumer, 1968), have been less successful and difficult to apply.

In most simple cases (unconstrained), random search, when far from the optimum, results in a probability of .5 for a successful move (Matyas, 1965). The effect of dimensionality on the efficiency of random and other search techniques has been studied by Schumer (1968), Lawrence (1973) and Anderssen (1978). In all cases, the number of function evaluations for convergence by random search was found to increase linearly with dimension, while all other techniques increased at a rate of higher order. In addition, while the rate of convergence is slowed, random search techniques were able to follow constraints, which caused failure in other methods. Therefore, the heavily constrained, high dimensional chemical process optimization problems should be an appropriate application for random search procedures, as has been the experience with the ARS.

ADAPTIVE RANDOMLY DIRECTED SEARCH

A new algorithm, termed the Adaptive Randomly Directed Search, ARDS, has been developed which incorporates several new features. This technique is similar to the Adaptive Random Search and determines new values of the independent variables from the following equation:

$$x_i = x_i^* + \frac{R_i}{K} \frac{(2\theta_i - 1)}{D} \quad (2)$$

$$D = \left[\sum_{i=1}^{N} (2\theta_i - 1)^2\right]^{1/2}$$

K = range reduction coefficient, a positive integer ($K=1,2,...$)

This technique randomly searches for improvements in the objective function from the surface of an n-dimensional ellipsoid centered about the best search point. A geometrical description of the search region in two dimensions is given in Figure 1. The axes of the n-dimensional ellipsoid are the ranges ($\frac{R_i}{K}$) of the independent variables. The quantity D reduces the size of the random step to unit length and the sampling region is restricted to the surface of the ellipsoid.

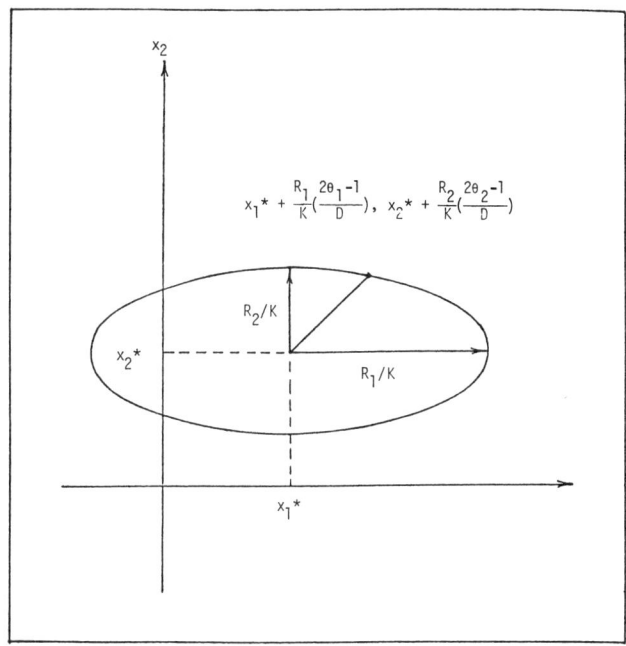

Figure 1. Geometrical Representation of the Adaptive Randomly Directed Search

The ranges, and thus the size of the ellipsoid, are decreased as the search converges to maintain a satisfactory ratio of improved search points to function evaluations. This means of range reduction allows the algorithm to use the history of the search to establish the forthcoming step size. Thus, a large step in one dimension and a small step in another may be dictated by the pattern of successful prior steps.

The reduction of the range is controlled by incrementing the range reduction coefficient K. K is increased to insure that the ratio of successful search points to function evaluations is maintained above a minimum value. The initial range, R_i, is set at a value less than the bounds of each independent variable. Periodically, the range is expanded if the algorithm stalls. The algorithm also allows the step size to be expanded once a successful direction has been established. The technique also allows movement opposite to an unsuccessful direction, since the probability of this direction improving the objective function is greater than another random move.

Application of ARDS removes the necessity for the user to make _a priori_ judgments concerning the solution of specific problems. The technique is easy to program and applicable to a wide variety of problems, including mixed-integer problems. The technique has been successfully applied to a number of example problems.

Example Problems

Six test functions were used to evaluate the capabilities of the proposed random search algorithm. A description of each is given below along with a brief summary of the other techniques used to solve each problem. In addition, any special techniques used in performing the optimization are outlined from the original literature source.

Problem 1. This example is a mathematical problem with a known optimum of -44 at values of the four independent variables of $x_1 = 0$, $x_2 = 1$, $x_3 = 2$, $x_4 = -1$. The problem was originally proposed by Rosen and Suzuki (1965) as a test for nonlinear programming algorithms. The problem was also solved by Gould (1971), who applied a modified sequential unconstrained technique to achieve the optimum. Luus and Jaakola (1973) and Heuckroth and Gaddy (1976) also solved the problem with random search procedures. The problem is stated as:

(minimize) (3)
$$y = x_1^2 + x_2^2 + 2x_3^2 + x_4^2 - 5x_1 - 5x_2 - 21x_3 + 7x_4$$

subject to:
$$x_1^2 + x_2^2 + x_3^2 + x_4^2 + x_1 - x_2 + x_3 - x_4 - 8 \leq 0$$
$$x_1^2 + 2x_2^2 + x_3^2 + 2x_4^2 - x_1 - x_4 - 10 \leq 0$$
$$2x_1^2 + x_2^2 + x_3^2 + 2x_1 - x_2 - x_4 - 5 \leq 0.$$

Problem 2. This problem, an alkylation process, was proposed by Payne (1958) and optimized by Sauer, et al, (1964) utilizing a linear programming technique with automatic approximation of the nonlinear model with small linear segments. Keefer (1973) also solved this problem with the Simpat method. The problem was also solved by random search techniques by Luus and Jaakola (1973) and Heuckroth and Gaddy (1976). The objective was to maximize the profit, Y, given by:

$$Y = .063 x_4 x_7 - 5.04 x_1 - .035 x_2 - 10 x_3 - 3.36 x_5$$

subject to: (4)

$$x_2 = x_1 x_8 - x_5$$
$$x_3 = 0.001 \, x_4 \, x_6 \, x_9 / (98 - x_6)$$
$$x_4 = x_1 (1.12 + .13167 x_8 - .00667 x_8^2)$$
$$x_5 = 1.22 x_4 - x_1$$
$$x_6 = 89 + (x_7 - (86.35 + 1.098 x_8 - .038 x_8^2))/.325$$

$$x_9 = 35.82 - .222 x_{10}$$
$$x_{10} = -133 + 3 x_7$$
$$0.01 \leq x_1 \leq 2000$$
$$0.01 \leq x_2 \leq 16{,}000$$
$$0.01 \leq x_3 \leq 120$$
$$0.01 \leq x_4 \leq 5000$$
$$0.01 \leq x_5 \leq 2000$$
$$85 \leq x_6 \leq 93$$
$$90 \leq x_7 \leq 95$$
$$3 \leq x_8 \leq 12$$
$$1.2 \leq x_9 \leq 4$$
$$145 \leq x_{10} \leq 162$$

where:

x_1 = olefin feed, barrels/day
x_2 = isobutane recycle, barrels/day
x_3 = acid addition rate, thousand pounds/day
x_4 = alkylate yield, barrels/day
x_5 = isobutane makeup, barrels/day
x_6 = acid strength, weight %
x_7 = motor octane number
x_8 = external isobutane-to-olefin ratio
x_9 = acid dilution factor
x_{10} = F-4 performance number

From the equality constraints, seven variables are eliminated and the search is performed over the independent variables x_1, x_7, and x_8. The optimum value of Y is 1162 at $x_1 = 1728.4$, $x_2 = 94.2$ and $x_3 = 10.4$.

Problem 3. This problem involves the maximization of the drying rate for a through-circulation dryer. The process was proposed by Chung (1972), who solved the system analytically using a differential algorithm. Luus and Jaakola (1973) and Heuckroth and Gaddy (1976) also optimized the process using random search techniques. The equations describing the process were formulated as follows:

(maximize) $y = .033 x_1 / H$ (5)

where:
$$H = .036/F + .095 - 9.27(10^{-4})/E(\ln(G/F))$$

$G = 1. - \exp(-5.39E)$

$F = 1. - \exp(-107.9E)$

$E = x_2/x_1^{.41}$

subject to:

$.2 - 4.62(10^{-10}) x_1^{2.85} x_2 - 1.055(10^{-4}) x_1 \geq 0$

$4/12 - 8.2(10^{-7}) x_1^{1.85} x_2 - 2.25/12 \geq 0$

$2. - 109.6(E)(H) \geq 0.$

A maximum value of $y = 172.5$ occurrs at $x_1 = 977.12$ and $x_2 = .523$.

<u>Problem 4</u>. The example is an algebraic test function proposed by Rosenbrock and Storey (1966). Malik and Hughes (1979) optimized this function with the Golden Complex Search. The optimum was easily found analytically to be $x_1 = 5, x_2 = 5$, with the optimum value of the objective function equal to zero. The problem was to minimize:

$$y = (x_1 - x_2)^2 + ((x_1 + x_2 - 10)/3)^2 \quad (6)$$

<u>Problem 5</u>. This problem was originally presented by Hesse (1973). Luus and Wang (1978) optimized the function using a new random search technique. The objective is to maximize:

$$y = 25(x_1 - 2)^2 + (x_2 - 2)^2 + (x_3 - 1)^2 + (x_4 - 1)^2$$
$$+ (x_5 - 1)^2 + (x_6 - 4)^2 \quad (7)$$

subject to:

$2 \leq x_1 + x_2 \leq 6$

$-x_1 + x_2 \leq 2$

$x_1 - 3x_2 \leq 2$

$(x_3 - 3)^2 + x_4 \geq 4$

$(x_5 - 3)^2 + x_6 \geq 4$

$0 \leq x_1$

$0 \leq x_2$

$1 \leq x_3 \leq 5$

$0 \leq x_4 \leq 6$

$1 \leq x_5 \leq 5$

$0 \leq x_6 \leq 10$

This problem had eighteen local maxima with the global maximum located at $x_i = 5,1,5,0,5,10$; $i = 1-6$, with $y = 310$.

<u>Problem 6</u>. This example is a geometrical problem presented by Luus (1974) and solved by Luus and Wang (1978) using a new random search technique. The objective is to find the maximum of the function:

$$y = x_1^2 + x_2^2 + x_3^2 \quad (8)$$

subject to:

$4(x_1 - .5)^2 + 2(x_2 - .2)^2 + x_3^2 + .1 x_1 x_2 + .2 x_2 x_3 \leq 16$

$2x_1^2 + x_2^2 - 2x_3^2 \geq 2$

$-2.3 \leq x_i \leq 2.7 \quad i=1,2.$

The optimum value of y is 11.68 at $x_1 = .989$, $x_2 = 2.674$ and $x_3 = -1.884$. This problem was proposed by Luus (1974) to illustrate the two-pass method for solving optimization problems containing difficult equality constraints. The two-pass method (Luus and Wang, 1978) was also used to solve the problem in this study. This technique first uses each equality constraint, $g_i(\underline{x})=0$, as two constraints $g_i(\underline{x}) < \alpha$ and $g_i(\underline{x}) > -\alpha$. The optimization is run to determine which of the two constraints is active. A second optimization with $\alpha=0$ for the active constraint yields the true optimum. As suggested by Luus (1974), the value of α for all constraints in this problem was set equal to one.

Results of the Solution of the Example Problems

Table 1 lists the efficiencies reported in the literature for each of the example problems. These problems are considered difficult in that irregular surfaces are involved with a number of implicit constraints, as shown by equations (3) - (8). The number of function evaluations required for convergence is used as the measure of efficiency, since these values are commonly reported and provide a direct comparison of the amount of computer time utilized. Table 1 shows that a large number of function evaluations is required for convergence of each of these problems.

Comparison with random search techniques is difficult since a random search is a stochastic process. Different search histories can be obtained from the same starting point, depending upon the sequence of random numbers used. Therefore, random search efficiencies should be based upon the average of a number of trials. As noted in Table 1, the average efficiency is reported for only some of the examples. Where comparisons are available, random search efficiencies show substantial improvement over other techniques.

ARDS was applied to these examples to determine the best values of the parameters. In each case, the examples were calculated in the same manner as the original literature source.

Starting points were chosen to be the same as for other solutions. To average the efficiency for solution with ARDS, 100 different seed numbers were used for the random number generator, and the average number of function evaluations for these 100 runs is reported.

Table 1
Average and Minimum Efficiencies Reported for the Example Problems

Test Prob.	Function Evaluations Required to Reach .1% of the Optimum		Literature Source
	Average	Minimum	
1	1948	190	Heuckroth & Gaddy, 1976
		1759	Luus & Jaakola, 1973a
		912	Gould, 1971
2	462	132	Heuckroth & Gaddy, 1976
		919	Luus & Jaakola, 1973a
		3487	Keefer, 1973
3	484	91	Heuckroth & Gaddy, 1976
		336	Luus & Jaakola, 1973a
4		152	Malik & Hughes, 1979
5		1674*	Luus & Wang, 1978
		1201**	Luus & Wang, 1978
		1787***	Luus & Wang, 1978
6		2175*	Luus & Wang, 1978
		3571**	Luus & Wang, 1978
		1942***	Luus & Wang, 1978

* With the method of Luus & Jaakola, 1973a
** With the method of Gaines & Gaddy, 1976
*** With the method of Luus & Wang, 1978

Parameter Selection

The parameters to be established for use with ARDS were the method for incrementing the range reduction coefficient, a means for setting the initial range of the variables and procedures for resetting the range.

Examples 2 and 3 were used to study various success ratios for controlling the range. The success ratio is defined as the number of successful function evaluations divided by the total number of function evaluations. If the cumulative ratio drops below the specified value, the range reduction coefficient is increased. Table 2 gives the average of the experiments conducted with various success ratios for a number of initial ranges. For Problem 2, the best ratio was found to be .5. However, a ratio of .25 gave consistently better results for Problem 3. Matyas (1965) showed that the success ratio for a random search should be .5 when the search is far from the optimum. For heavily constrained problems, a search ratio of less than .5 would be expected. A success ratio of .4 was chosen as appropriate for constrained problems and was used in ARDS.

A series of studies was made with Problem 1, 2, and 3 to determine the best means of increasing the range reduction coefficient. Increments of .25, .5, .75, and 1.0 were examined and the results are shown in Table 3. An increment of .25 was found to result in nonconvergence in most cases. An increment of 1.0 was found to give the highest convergence reliability, but not always the best search efficiency. Therefore, the range in ARDS is controlled by incrementing the range reduction coefficient by 1 when the cumulative success ratio drops below .4.

Table 2
Results of Varying Success Ratio of ARDS

Success Ratio	Function Evaluations Required to Reach .1% of Optimum	
	Problem 2	Problem 3
0.0	NC	259
0.25	173	106
0.5	135	195
0.75	139	329
1.0	161	564

Table 3
Results of Range Reduction of ARDS

Increment of K	Function Evaluations Required to Reach .1% of Optimum		
	Problem		
	1	2	3
0.25	NC	NC	79
0.5	156	110	70
0.75	149	77	61
1.0	160	126	54

The initial range of the variables was found to be an important parameter in achieving good efficiency. The best initial range was found to be only a fraction of the available range of each variable. Multiples of the order of magnitude of the starting value of the independent variables x_i's were studied for the example problems. The order of magnitude is defined as $10^{int(\log x_i)}$. The initial values of the independent variables are specified by the user and may be in the neighborhood of the optimum where only small step sizes would be required. The use of small initial ranges improves the reliability but may impair the efficiency of the search. Differences between the initial values and the upper or lower bound were also tried and found to be less successful.

Table 4 gives the results of varying the initial ranges for Problems 1, 2, and 3. Higher multiples were also used and found to be unsuccessful. The best results were obtained by the use of 2 times the order of magnitude of the initial value of each independent variable. The highest reliability was also obtained with this initial range. Ratios of less than one were also studied and found to be successful in many cases. Therefore, the algorithm is also programmed such that the initial range is halved if no improvements are found in the first five function evaluations. Thus, smaller ranges will be quickly established if larger step sizes are not appropriate.

If the program does not make adequate progress toward convergence, the algorithm will reset the ranges to facilitate faster movement. At this point, the counters for the success ratio are reset to their initial values, and the ranges are reduced to a fraction of the initial value. The search is then continued from the best search point which has been found. Various divisors for resetting the ranges were evaluated on the test problems. These results are also shown in Table 4. As noted, the series of divisors (5, 10, 15, 20, ...) was found to give the best convergence reliability and an acceptable efficiency.

The initiation of the restart procedure was also a variable studied. Changes in the independent variables were used as the criteria for restarting the search. If the movement of the search slows, such that the independent variables change only a small amount, it is necessary to reset the ranges before continuing the search. Changes of a minimum of one percent and .1 percent of the initial step size of each variable were studied. Nearly identical efficiencies were yielded for all the test problems with both these restart criteria. Therefore, the smaller percentage was chosen. The algorithm restarts the search if a change in the step size is less than .1 percent after 5 improvements in the objective function.

Comparison of Efficiencies. A comparison of the efficiency of ARDS with other techniques for all the example problems is given in Table 5. The average and minimum number of function evaluations for ARDS is shown, as are the minimum values reported in the literature for the other methods. ARDS demonstrated a substantial improvement in efficiency in all cases. The average efficiency is up to 80 percent better than the minimum efficiency reported for other procedures. In addition, the dependability of convergence of ARDS was greater than 90 percent in nearly all cases. The dependability for the ARS for a somewhat simpler group of problems was 80 percent (Heuckroth, 1975).

Table 4
Results for Varying the Initial Range of Variables in ARDS

Series of Divisors for the Initial Range for Each Restart	Function Evaluations Required to Reach 0.1% of the Optimum								
	Test Problem 1			Test Problem 2			Test Problem 3		
	$R_i = y^*$	$R_i = 2y$	$R_i = 3y$	$R_i = y$	$R_i = 2y$	$R_i = 3y$	$R_i = y$	$R_i = 2y$	$R_i = 3y$
(2,3,4,...)	190	NC	93	84	NC	NC	71	49	105
(3,5,7,...)	172	NC	131	128	91	NC	106	62	92
(4,7,10...)	188	NC	127	98	111	NC	74	49	112
(5,10,15,...)	190	155	94	103	101	117	71	54	106
(10,20,30,...)	193			102			54		

*Order of Magnitude of the starting value

Table 5
Comparison of Search Efficiencies

Example Problem	Function Evaluations Required to Reach .1% of Optimum		Minimum Function Evaluations From Literature
	Average	Minimum	
1	155	39	190 (ARS)
2	101	36	132 (ARS)
3	54	5	91 (ARS)
4	92	29	152 (Malik & Hughes, 1974)
5*	899	407	1201 (ARS)
6**	363	43	1942 (Luus & Wang, 1978)

* Average for 12 starting points
** Average for 7 starting points

The automatic procedures for setting the initial ranges worked well for Examples 1-3. However, for Problems 4, 5 and 6 larger ranges gave better efficiencies in converging to .1% of the optimum. The best initial variable range is dependent upon the functional surface, the constraints and the starting point. Since these considerations will not be known in advance, the selection of the initial range may best be made by the user, after some preliminary esperimentation. the success ratio and range reduction criteria and restart procedures worked well for all examples and, probably, can be successfully applied to most problems. It should be noted that these examples are heavily constrained with active constraints at or near the optimum. ARDS was found to effectively follow the constraints and efficiently find the solution without modification.

CONCLUSIONS

The new algorithm, ARDS, has been demonstrated to be efficient in solving several difficult (heavily constrained) optimization problems from the literature. This algorithm results in up to an 80 percent reduction in the number of function evaluations to converge to the optimum and exhibits a reliability of over 90 percent.

General procedures for utilizing the search history to adjust the search region were developed and implemented. A procedure for selecting the initial ranges were also developed, but was not entirely successful in all problems. Thus, the use of the algorithm required specification of only the starting point and, in some cases, the initial ranges.

LITERATURE CITED

1. Anderssen, R. S. and P. Bloomfield, "Properties of the Random Search in Global Optimization," Journal of Optimization Theory and Applications, Vol. 16, No. 5/6, 383 (1975).

2. Baba, N. and T. Shoman, "A Modified Convergence Theorem for a Random Optimization Method," Information Sciences, Vol. 13, 159 (1977).

3. Ballman S., and J. L. Gaddy, "Optimization of Methanol Process by Flowsheet Simulator," Ind. Eng. Chem. Process Des. Dev., Vol. 16, No. 3, 337 (1977).

4. Barneson, R. A., N. F. Brannock, J. G. Moore, and C. Morris, Chem. Eng., (July 27, 1970).

5. Bekey, G. A., M. H. Gran, A. E. Sabroff, and A. Wong, "Parameter Simulation by Random Search Using Hybrid Computer Techniques," AFIPS Conference Proc., Vol. 29 (1966).

6. Braken, J. and G. P. McCormick, Selected Applications of Nonlinear Programming, J. Wiley, New York (1968).

7. Brooks, S. H. "A Discussion of Random Methods for Seeking Maxima," Oper. Res., Vol. 6, 244, (1958).

8. Chung, S. F., "Mathematical Model and Optimization of Drying Process for a Through-Circulation Dryer," Can. J. Chem. Engr., Vol. 50, 657 (1972).

9. Doering, F. J. and J. L. Gaddy, "Optimization of the Sulfuric Acid Process with a Flowsheet Simulator," Computers and Chemical Engineering, Vol. 4, No. 2, 113 (1980).

10. Friedman, P. and K. L. Pinder, Ind. Eng. Chem. Proc. Des. Develop., Vol. 11, 512 (1972).

11. Gaines, L. D. and J. L. Gaddy, "Process Optimization by Flowsheet Simulation," Ind. Eng. Chem. Process Design Develop., Vol. 15, No. 1, 206 (1976).

12. Gall, D. A., "A Practical Multifactor Optimization Criterion," Recent Advances in Optimization Techniques, A. Lavi and T. P. Vogl, ed, Wiley, New York (1966).

13. Gottfried, B. S., P. R. Bruggink, and E. R. Harwood, "Chemical Process Optimization Using Penalty Functions," Ind. Eng. Chem. Process Design Develop., Vol. 9, 581 (1970).

14. Gould, F. J., "Nonlinear Pricing: Applications to Concave Programming," Operations Research, Vol. 19, 1026 (1971).

15. Hesse, R., "A Heuristic Search Procedure for Estimating a Global Solution of Nonconvex Programming Problems," Opns. Res., Vol. 21, 1267 (1973).

16. Heuckroth, M. W., L. D. Gaines, and J. L. Gaddy, "An Examination of the Adaptive Random Search Technique," AIChE Journal, Vol. 22, No. 4, 744 (1976).

17. Jinkerson, K. R. and J. L. Gaddy, "Ethylene Process Optimization. Constraint Relaxation and Bounds Adjustment," Ind. Eng. Chem. Process Des. Dev., Vol. 18, No. 4, 579 (1979).

18. Keefer, D. L., "Simpat: Self-Bounding Direct Search Method for Optimization," Ind. Eng. Chem. Process Design Develop., Vol. 12, 92 (1973a).

19. Kelahan, R. C. and J. L. Gaddy, "Application of the Adaptive Random search to Discrete and Mixed Integer Optimization," *International Journal for Numerical Methods in Engineering*, Vol. 12, 289 (1978).

20. Lawrence III, J. P. and K. Steiglitz, "Randomized Pattern Search," *IEEE Transactions on Computers*, Vol. C-21, No. 4, 382 (1972).

21. Luus, R., "Two-Pass Method for Handling Difficult Equality Constraints in Optimization," *AIChE J.*, Vol. 20, 608 (1974).

22. Luus, R., and T.H.I. Jaakola, "Optimization by Direct Search and Systematic Reduction of the Size of Search Region," *AIChE J.*, Vol. 19, 760 (1973a).

23. Malik, R. K. and R. R. Hughes, "Optimization of Noisy Functions," paper presented at AIChE meeting, Houston, Texas (Apr. 1979).

24. Matyas, J., "Random Optimization," *Automatic and Remote Control*, Vol. 26, 246 (1965).

25. Payne, R. E., "Alkylation-What You Should Know About This Process," *Petrol. Refiner*, Vol. 37, 316 (1958).

26. Price, W. L., "A Controlled Random Search Procedure for Global Optimization," *Computer Journal*, Vol. 20, No. 4, 367 (1977).

27. Rosenbrock, H. H. and C. Storey, *Computational Techniques for Chemical Engineers*, Pergamon Press, Oxford, (1966).

28. Rosen, J. B. and S. Suzuki, "Construction of Nonlinear Programming Test Problems," *Communications of the ACM*, Vol. 8, 113 (1965).

29. Sauer, R. N., A. R. Colville, and C. W. Burwick, "Computer Points in the Way to More Profits," *Hydrocarbon Processing Petrol. Refiner*, Vol. 43, 84 (1964).

30. Schumer, M. A. and K. Steiglitz, "Adaptive Step Size Random Search," *IEEE Transactions on Automatic Control*, Vol. AC-13, No. 3, 270 (1968).

31. Seader, J. D. and D. E. Dallin, *Chem. Eng. Computing*, Vol. 1, AIChE Workshop Series (1972).

32. Shannon, P. T., A. I. Johnson, C. M. Crowe, T. W. Hoffman, A. E. Hamielec, and D. R. Woods, *Chem. Eng. Prog.*, Vol. 62.6, 49 (1966).

33. Sprang III, H. A., "A Review of Minimization Techniques for Nonlinear Functions," *SIAM Review*, Vol. 4, No. 4 (1962).

34. Umeda, T., and A. Ichikowa, "A Modified Complex Method for Optimization," *Ind. Eng. Chem. Process Design Develop.*, Vol. 10, 229 (1971).

35. Wang, B. C. and R. Luus, "Reliability of Optimization Procedures for Obtaining Global Optimum," *AIChE J.*, Vol. 24, 619 (1978).

36. Westerberg, A. W., T. J. Berna, and M. H. Locke, "A New Approach to Optimization of Chemical Processes," *AIChE J.*, Vol. 26, No. 1, 37 (1980).

37. Wozny, M. J. and G. T. Heydt, "Hyperconical Random Search," *Transactions of the ASME*, Series G., Vol. 94, No. 1, 71 (1972).

38. Zellnik, H. E., N. E. Sondak, and R. S. Davis, *Chem. Eng. Prog.*, Vol. 58, No. 8 (1962).

APPLICATION OF DACSL (DOW ADVANCED CONTINUOUS SIMULATION LANGUAGE) TO THE DESIGN AND ANALYSIS OF CHEMICAL REACTOR SYSTEMS

GARY L. AGIN
and
GARY E. BLAU

Systems Research Laboratory
The Dow Chemical Company
Midland, MI 48640

The Dow Advanced Continuous Simulation Language (DACSL) is a software package for modeling and analyzing the dynamic behavior of physical phenomena. It is designed to meet the needs of scientists and engineers who wish to model systems without having to spend the time and resources to learn details of sophisticated computer programming. The DACSL package provides a simple method of programming the mathematical models which are typically characterized by time-dependent nonlinear differential equations. The package is capable of performing parameter estimation based on experimental data, and process optimization. It displays results in graphical and tabular forms. DACSL is not available in the public domain. However, it consists of several software packages which are available from various vendors. Various features of DACSL are illustrated with an example of ethanolamine reaction kinetics.

In the last few years the chemical processing industry has been stimulated to build new plants which are both energy efficient and minimize the production of any environmentally unattractive by-products. Perhaps the major factor influencing plant design is the optimal design and operation of the chemical reactor itself. By understanding the chemistry and kinetics of the reaction system, it is frequently possible to modify the design of the entire plant or design a new reactor to minimize energy intensive recycle streams and/or environmentally unattractive by-products.

An inherent part of the reactor design problem is developing a mathematical model of the reactor system. In industrial practice this is frequently an interdisciplinary effort involving an experimentalist (frequently a chemist) and an engineer working in an interactive experimentation-analysis fashion. One or more candidate models are proposed to describe the reaction system. Based on these models experiments are designed and conducted to discriminate between these rival models. The experimental data are then analyzed to determine the most likely model. If this model is still inadequate, new models are suggested and the process is repeated. Kittrell (1970) has described this approach in detail.

In most cases the model equations, which arise from material and energy balances across various reactor geometries, i.e., tubular or stirred tanks, are systems of nonlinear ordinary differential equations with known initial conditions. Phenomenological kinetic parameters appearing in these equations are rarely known a priori and must be estimated from the experimental data for each model. Consequently, it is necessary to imbed numerical ordinary differential equation solving routines within a nonlinear parameter estimation algorithm. The numerical analysis problem is further complicated because the differential equation systems are frequently stiff. In stiff systems fast and slow mechanisms operate together making the numerical solution of the equations very difficult. And finally, there is an incentive to have this modeling work carried out by analysts who are not familiar with the fundamentals of solving differential equations or estimating parameters. That is, the modeling system needs to be user friendly.

In recent years several computer languages have been developed for representing and solving ordinary differential equation systems characterizing

---

G. E. Blau is now with Carnegie-Mellon University, Pittsburgh, Pennsylvania

0065-8812-82-6270-0214-$2.00.
© The American Institute of Chemical Engineers, 1982

various physical processes. The most widely used of these languages are ACSL, CSSL IV and CSMP III. Each of these languages was developed primarily to model electrical, mechanical and aerospace systems. These systems are characterized by a well defined structure which arise from electric circuit theory or basic mechanical principles so that most of the parameters such as resistances, capacitances, spring constants, gravitational constants are known. In modeling these systems it is only necessary to simulate, i.e. solve the differential equation system for known initial conditions and parameter values, to study the behavior of the physical process as a function of time. Although ACSL and CSSL IV are designed to solve stiff systems they are not designed to estimate model parameters from experimental data.

This paper describes a simulation package developed at The Dow Chemical Company for modeling not only chemical reactor systems but also general physical systems in which there is a need to estimate some of the model parameters from experimental data. This package has had a significant impact upon the process modeling community by dramatically increasing the productivity of experienced model builders and by placing a powerful tool into the hands of the non-computer oriented analyst. Although the package is not available in the public domain, it consists of existing programs which are available and could be unified into an analogous modeling and analysis system.

DACSL SYSTEM OVERVIEW

Simulation Language

The simulation language selected for defining the math models was the Advanced Continuous Simulation Language (ACSL) from Mitchell and Gauthier Associates, Inc. of Concord Mass. This language was selected over CSSL IV and CSMP III. The latter was not used because IBM has stopped supporting it and its integration methods do only a fair job of solving stiff systems. Also CSMP III was not designed with interactive time-sharing execution capabilities. ACSL was selected over CSSL IV because of favorable user comments and because FORTRAN source code was supplied with the system. Source code was required to permit extensive modification of the existing simulation language to meet Dow's needs.

ACSL is a FORTRAN based simulation language that was designed for scientists and engineers, who wish to study the dynamic behavior of physical phenomena without having to spend time and resources to learn sophisticated computer programming details. It permits them to concentrate on the problem being solved and not on the numerical methods involved in its solution. The program allows models to be prepared directly from sets of algebraic and ordinary differential equations. The program solves these equations using state-of-the-art numerical analysis methods, including Gear's method for stiff systems. The language is easy to use for the novice and yet contains a powerful macro language and permits use of FORTRAN for the more experienced user. The ACSL language was not modified for use in DACSL. The ACSL statements follow the standard of the Continuous Systems Simulation Language (CSSL). Committee of the Society for Computer Simulation released in December, 1967.

ACSL has a command driven run-time environment that permits the user to interact with the model while studying its behavior. In this environment the program accepts commands to change current values of model data (SET), to display model data (DISPLAY), to start a simulation run (START), print results in tabular form (PRINT), and to plot the results in graphical form (PLOT). This phase of the program was modified to form DACSL. Commands were added to specify parameters to be estimated (VARY), to specify data to be used in estimation (DATA and FIT), to specify process variables to be optimized (MINIMIZE and MAXIMIZE), and to start the optimization (ESTIMATE). Also, changes were made to provide on-line HELP information and more plotting options.

Nonlinear Programming Logic

Nonlinear programming software is needed in DACSL to estimate parameters and optimize parameters in process models. In the last few years significant progress has been made in developing new algorithms, modifying existing ones and in developing good software. (Lasdon (1980) has written an excellent survey of the latest methods and their performances on various test problems). One of the most effective algorithms has been the generalized reduced gradient (GRG) method. A brief description of the method appears in Lasdon (1980), but more detailed descriptions are available in Sargent (1979). A version of the GRG method has been implemented in the code GRG2 by Lasdon et al. (1978). This code has been incorporated into DACSL to perform the required parameter estimation. In the case where there are no design constraints on the problem, GRG2 code reduces to the powerful variable metric method described by Bard (1974). The advantage of these methods is that they develop an approximation to the covariance matrix of the optimal parameter estimates $\bar{\theta}$, an invaluable asset in evaluating the statistical quality of these best parameter estimates.

There is no guarantee of convergence to a global optimum from some initial user supplied estimate θ^0. Consequently, it is necessary for the user to make an effort to determine reasonable approximations to the optimal parameters $\bar{\theta}$. Failure to do so may not only result in converging to a false optimum but also may result in using an inordinate amount of computation time. A feature of this program is its interactive nature of the program to aid in finding initial parameter estimates and to follow the progress of the optimization. Graphical run time features have been built into DACSL to monitor the progress of the GRG2 algorithm well as the facility to readily change or update the parameter estimates θ^*. In practice one way to assure that the best estimates have been achieved is to start the problem from different points $\underline{\theta^0}$ and make sure they all converge to the same point $\underline{\theta^*}$.

Once the best estimates $\underline{\theta^*}$ have been determined DACSL prints out sufficient statistical information to test the adequacy of the model. A comparison between the measured and calculated values as well as the various dependent variables can be observed by invoking various DACSL graphic features.

If several candidate models are to be tested, the model equations are readily changed in DACSL and the process repeated.

Parameter Estimation Objective Function

The nonlinear parameter estimation problem is to find those values of the parameters θ^* which "best" describe the experimental data while simultaneously satisfying the design relations. What is "best"? This can only be answered by examining the error structure of the data.

Let f_i represent the value of the math model at independent variable t_i at given values of the parameters $\bar{\theta}$. The difference between the measured values z_i and predicted values f_i are the residuals

$$e_i = z_i - f_i \qquad (1)$$

If the model equation adequately describes the data, these residuals are estimates of experimental error. Although it is dangerous to generalize, we have found that the following error structure is characteristic of many process modeling applications:

(i) the observation errors at different time points t_i are uncorrelated

(ii) the errors are normally distributed with zero means and covariance matrix $\underline{V}$

(iii) the measured variables are independent so that the covariance matrix $\underline{V}$ is diagonal,

(iv) The magnitude of v_i is proportional to the magnitude of the measured variable z_i. That is

$$v_i = \omega^2 f_i^\gamma \qquad (2)$$

When $\gamma = 0$, the absolute error is constant throughout the experiment; when $\gamma = 2$, the relative error is constant throughout the experiment.

Reilly et al. (1977) have shown this heteroscedasticity parameter γ, varies between 0 and 2 although applications exist where $\gamma > 2$. This corresponds to the physical situation in which the relative error decreases with the magnitude of the relative error. In any case the parameter γ and ω^2 are generally unknown and must be estimated with the model parameters $\underline{\theta}$.

Since the form of the probability distribution of the errors is known, the most useful tool for parameter estimation is the maximum likelihood principle (Bard, 1974). In this approach, the parameters appearing in the probability distribution,

γ and ω^2 are estimated along with the unknown parameters in the model. Applying the maximum likelihood principle gives the following objective function which must be maximized.

$$L(\underline{\theta}, \underline{v}) = -2N \log 2\pi - \frac{1}{2} \sum_{i=1}^{N} \log V_i$$
$$- \frac{1}{2} \sum_{i=1}^{N} \frac{(z_i - f_i)^2}{V_i} \quad (3)$$

When V_i from Equation (2) is substituted into Equation (3), the resulting expression differentiated with respect to ω and set equal to zero, then solved

$$\omega^2 = \frac{1}{N} \sum_{i=1}^{N} \frac{(z_i - f_i)^2}{f_i^\gamma} \quad (4)$$

The following function results when ω is eliminated from the expression

$$L(\underline{\theta}, \gamma) = 2N(\log 2\pi + 1) - \frac{1}{2} \sum_{i=1}^{N} \gamma \log f_i$$
$$- N \log \frac{1}{N} \sum \frac{(z_i - f_i)^2}{f_i^\gamma} \quad (5)$$

The above development has been shown for a single observed variable vector. The overall objective function is represented as

$$L(\underline{\theta}, \underline{\gamma}) = \sum_{i=1}^{M} L_i(\underline{\theta}, \gamma_i) \quad (6)$$

where M represents the number of observed variables. The parameter estimation problem is to maximize this likelihood function to estimate $\underline{\theta}$ and $\underline{\gamma}$ subject to the design constraints. Equation (6) forms the objective function used in DACSL.

Addition of Nonlinear Programming to ACSL

The ACSL system translates the user model statements into FORTRAN subroutines, compiles them and calls these routines in a run-time command processor. This subroutine (ZZEXEC) reads the users' commands, interprets them, finds the command in a list of accepted keywords and then branches to other routines to process them. The list of accepted commands was enlarged to allow specification and solution of the nonlinear programming problem. Routines were written to examine the new commands and store the results in a labelled common area.

GRG2 is a modular program written as a collection of subroutines. Access to the logic is through a single subroutine (GRGITN) call. Most of the data needed by this routine is passed through an argument list. The GRG2 software calls a subroutine (GCOMP) that calculates the objective function at the current parameter values.

The writing of GCOMP Subroutine was the most difficult task within this work. A START command in ACSL calls the users model routine (ZZSIML) where initial conditions are calculated, model equations solved by numerical integration and data results reported at an interval specified by the user, and additional user calculations are performed at the end of the simulation. The structure of ZZSIML was modified by changing system macros that are invoked to generate this routine. These changes permits greater control of a simualtion calculations.

GCOMP then sets initial conditions from experimental data and integrates to the first independent variable value. The model values corresponding to experimental data values are stored and integration proceeds to the next data point. When the range of experimental data has been covered then the objective function is calculated.

The objective function is a function of the parameters, θ, alone. The optimum value of the function with respect to the heteroscedasticity parameter, γ, is found by a line search for each column of experimental data.

The approach used to implement GRG2 in to ACSL could and has been used to make available other optimization algorithms within DACSL. Most nonlinear programming routines are written in subroutine form and call routines that calculate objective functions.

Gradients required by GRG2 are calculated by making successive calls to GCOMP with the parameters being perturbed.

ILLUSTRATION OF DACSL FEATURES

More details on use of DACSL are illustrated by modeling the reaction of ethylene oxide (EO) and aqueous ammonia (NH_3) to form monoethanolamine (MEA), diethanolamine (DEA), and triethanolamine (TEA). The proposed reaction routes for these products are:

$$EO + NH_3 \xrightarrow{k_1} MEA$$
$$EO + MEA \xrightarrow{k_2} DEA$$
$$EO + DEA \xrightarrow{k_3} TEA$$

The reaction rate constants for the reactions are k_1, k_2, and k_3 respectively.

A math model was developed to estimate these reaction rate constants utilizing the published experimental data of Miki et al. (1966). These data values were obtained in a constant temperature batch reactor. The formation of the reaction products can be represented by the following ordinary differential equations,

$$\frac{d[MEA]}{dt} = k_1[NH_3][EO] - k_2[MEA][EO] \quad (7)$$

$$\frac{d[DEA]}{dt} = k_2[MEA][EO] - k_3[DEA][EO] \quad (8)$$

$$\frac{d[TEA]}{dt} = k_3[DEA][EO] \quad (9)$$

These equations assume a constant reactor contents volume and use brackets ([]) to indicate concentration values. The concentration profiles for EO and NH_3 can be found from material balances. To avoid computing negative EO values at high conversions due to round-off error, it is better to represent EO material balance with a differential equation in terms of ln[EO] as a variable Equation (10).

$$\frac{d(\ln[EO])}{dt} = -k_1[NH_3] - k_2[MEA] - k_3[DEA] \quad (10)$$

The experimental data consists of percentage EO conversion as a function of time and the final percentage distribution of EO in the products. The model must contain these quantities calculated from the composition profile values.

DACSL Simulation Language Features

Math models such as the ethanolamine reaction kinetics are defined within DACSL using the syntax of the ACSL simulation language. The language is structured to produce efficient execution by performing calculations only when needed. This structure is shown in Table 1.

TABLE 1 - DACSL Language Structure

```
PROGRAM simulation title
  INITIAL
     Initial conditions calculations
  END
  DYNAMIC
     DERIVATIVE
        Definition of dynamic model statements
     END
     Communication interval calculations
  END
  TERMINAL
     Calculations at end of the model
  END
END
```

The INITIAL section defines the initial conditions for the model variables and usually contains statements pertaining to model data. These statements are shown in Table 2. This section can also contain FORTRAN statements to calculate initial conditions from variables defined in CONSTANT statements.

TABLE 2 - Data Statements

INTEGER	Specifies integer variables
LOGICAL	Specifies logical variables
REAL	Specifies real variables
ARRAY	Defines the size of vectors and matrices of upto three dimensions
CONSTANT	Defines initial values of variables
EQUIVALENCE	Specifies variable mapping
TABLE	Defines a function generator of upto three dimensions

The DERIVATIVE section defines the dynamic model and contains statements representing the numerical integration operation. This operator is the key to this simulation language. Table 3 shows the comparison between mathematical representation of a differential equation and its corresponding DACSL statements.

TABLE 3

Comparison of DACSL with Math Notation

Mathematics	DACSL
$\frac{dy}{dt} = -k_1 y - k_2 y^2$	DYDT = -K1*Y-K2*Y*Y
$y = \int \frac{dy}{dt} dt + y(t=0)$	Y = INTEG(DYDT, Y0)

Table 4 shows some of the calculus operators available in DACSL. In addition, several forcing functions, random number generators, function switches, and all FORTRAN system subroutines are available to aid in defining the dynamic model. Implicit equations in a single variable can also be solved within the DACSL language.

TABLE 4 - DACSL Calculus Operators

INTEG	Integration of a single variable
INTVC	Integration of a vector or matrix
LIMINT	Limited integrator
DBLINT	Double limited integrator
MODINT	Moded integrator
INTGRL	CSMP III integrator
DERIVT	Derivative of a function
REALPL	First-order transfer function
TRAN	General transfer function
DEAD	Dead space
DELAY	Transportation lag
LEDLAG	Lead-lag compensator

DACSL also contains a powerful macro language that allows the modeller to define new operators and to ease program development. A feature that generates new variable names by concatenation is especially useful. The simulation program can also be programmed in FORTRAN subroutines and called from the DACSL program.

A DACSL program representing the ethanolamine reaction kinetics is shown in Figure 1. Note that more than one statement can be on a line through uses of a dollar sign ($), statements are continued using an ellipsis (...), and comments included in quotes (') are allowed. These comments document the purpose of each statement in the example. Also note that the equations follow the syntax of FORTRAN.

The INITIAL section of the model calculates the initial component concentrations from the feed, NH_3 to EO mole ratio, and weight percentage NH_3 in the

FIGURE 1 - DACSL Example Program

```
PROGRAM - ETHANOLAMINE REACTION KINETICS EXAMPLE
INITIAL
VARIABLE TIME  = 0.                $'RENAME TIME VARIABLE'
CONSTANT MWH2O =  18.02, MWNH3 =  17.03, ...
         MWEO  =  44.05, MWMEA =  61.08, ...
         MWDEA = 105.14, MWTEA = 149.19  $'MOLE WEIGHTS'
CONSTANT K1=.005, K21=1., K31=1.   $'RATE CONSTANTS'
CONSTANT MEAIC=0., DEAIC=0., TEAIC=0.   $'PRODUCTS WEIGHTS'
CONSTANT RATIO = 5.                $'NH3/EO MOLE RATIO'
CONSTANT NH3AQ = 18.               $'WT % NH3 IN WATER'
CONSTANT TEMP  = 30.               $'TEMPERATURE (DEG C)'
CONSTANT TSTOP = 300.              $'REACTION TIME (MIN)'
CINT     = 0.01 * TSTOP            $'COMMUNICATION INTERVAL'
'CALCULATE MOLES OF REACTANTS BASED ON EO'
EOIC   = 1.                        $'INITIAL MOLES EO'
NH3IC  = RATIO * EOIC              $'INITIAL MOLES NH3'
WTNH3  = NH3IC * MWNH3             $'WT NH3 FOR NEXT CALC'
WTH2O  = 100. * WTNH3 / NH3AQ - WTNH3  $'WT H2O (CONSTANT)'
H2OIC  = WTH2O / MWH2O             $'MOLES H2O (CONSTANT)'
'CALCULATE MOLAR CONCENTRATIONS - MOLES / LITER'
VOLUME = VOL( H2OIC, NH3IC, EOIC, MEAIC, DEAIC, TEAIC, TEMP )
H2OIC  = H2OIC / VOLUME            $'INITIAL CONC. H2O'
NH3IC  = NH3IC / VOLUME            $'INITIAL CONC. NH3'
EOIC   = EOIC  / VOLUME            $'INITIAL CONC. EO'
LNEOIC = ALOG( EOIC )              $'INITIAL LN(CONC. EO)'
MEAIC  = MEAIC / VOLUME            $'INITIAL CONC. MEA'
DEAIC  = DEAIC / VOLUME            $'INITIAL CONC. DEA'
TEAIC  = TEAIC / VOLUME            $'INITIAL CONC. TEA'
'CALCULATE REACTION RATE CONSTANTS'
K2     = K21 * K1                  $'K21 = K2 / K1'
K3     = K31 * K1                  $'K31 = K3 / K1'
END$'OF INITIAL'

DYNAMIC

DERIVATIVE

'CALCULATE NH3 FROM A NH3 BALANCE'
NH3    = NH3IC + MEAIC - MEA + DEAIC - DEA + TEAIC - TEA
'CALCULATE MOLAR RATES OF CHANGE - MOLES / MINUTE'
R1     = K1 * EO * NH3             $'NH3 TO MEA REACTION'
R2     = K2 * EO * MEA             $'MEA TO DEA REACTION'
R3     = K3 * EO * DEA             $'DEA TO TEA REACTION'
XLNEO  = - ( R1 + R2 + R3 ) / EO   $'LN(EO) LOSS'
XMEA   = R1 - R2                   $'MEA FORMATION'
XDEA   = R2 - R3                   $'DEA FORMATION'
XTEA   = R3                        $'TEA FORMATION'
'CALCULATE MOLES BY INTEGRATION OF DIFFERENTIAL EQUATIONS'
LNEO   = INTEG( XLNEO, LNEOIC )    $'CURRENT LN(EO) MOLES'
EO     = EXP( LNEO )               $'CURRENT EO MOLES'
MEA    = INTEG( XMEA,  MEAIC  )    $'CURRENT MEA MOLES'
DEA    = INTEG( XDEA,  DEAIC  )    $'CURRENT DEA MOLES'
TEA    = INTEG( XTEA,  TEAIC  )    $'CURRENT TEA MOLES'
TERMT (TIME .GE. TSTOP)            $'STOP IF TIME ¶ TSTOP'
END$'OF DERIVATIVE'
'CALCULATE VALUES FOR COMPARISON WITH DATA'
CNVNH3 = 100. * ( NH3IC - NH3 ) / NH3IC  $'% NH3 CONVERSION'
CNVEO  = 100. * ( EOIC  - EO  ) / EOIC   $'% EO CONVERSION'
IF ( ABS(CNVEO) .LT. 1.E-4 ) CNVEO = 0.
MON    = MEA + 1.E-9               $'AVOIDS ZERO DIVIDE'
EA     = MON + 2.*DEA + 3.*TEA     $'TOTAL EO IN PRODUCTS'
MONO   = 100. * MON / EA           $'% EO IN MEA'
DI     = 200. * DEA / EA           $'% EO IN DEA'
TRI    = 300. * TEA / EA           $'% EO IN TEA'
END$'OF DYNAMIC'
END$'OF PROGRAM'
```

aqueous ammonia feed. A function subprogram named VOL returns volume of the reaction ingredients needed in the molar concentration calculation. The reaction rates k_2 and k_3 are expressed as a function of k_1.

The dynamics of the system are described in the DYNAMIC section. Within this section is the DERIVATIVE section containing the equations that define the differential equations and sets up their solution. The NH_3 concentration is determined from a material balance. The reaction rate expressions are then calculated from the rate constants and concentration values. These expressions are used to find the derivatives of concentration values with respect to time. The structure of the differential equations is defined in the statements containing the INTEG integration operators.

The remaining statements within the DYNAMIC section, but outside the DERIVATIVE section are calculated only when model values are to be reported. These expressions are not necessary for the solution of the equations, but are values that must be calculated for comparison with experimental data. The values calculated are percentage conversion of NH_3 and EO and the percentage distribution of EO in the reaction products.

A final section called TERMINAL could be used to represent values that need only to be calculated once at the end of the simulation. In this example no calculations are required and the section was omitted.

Translator Features

The translator reads the DACSL language statements and converts them into FORTRAN instructions. In this process, it builds a dictionary of variable names, their type, and length and passes it as part of the generated FORTRAN code. As a result during the next run-time phase, variables can be referenced by name within commands.

The translator also sorts statements in the DERIVATIVE section into the proper execution order. This permits model definition statements to be in any order. The modeller can thus focus his attention on the problem and not on the calculation order required by the computer.

Reporting of errors is another function of the translator. It notes variables that were not assigned values, statements that could not be sorted, and other syntax errors. This data is quite helpful in locating programming errors.

Run-Time Features

The simulation package operating on a time-sharing computer permits the simulator to interact with the model to study its behavior. Commands are accepted to let the modeller change or display data values, start or continue a simulation, estimate model parameters and print or plot simulation results. Table 5 shows the run-time commands that are currently available in DACSL. The commands that were added to the ACSL command processor are primarily for specifying and solving parameter estimation and process optimization problems. The PLOT command was modified to permit logrithmic axis types, plotting of experimental data values, renaming the axis labels, generating grids, and plotting of curves in color on terminals that support it.

Simulation Run-Time Command Example. Figure 2 shows a listing that contains some of the run-time commands available in DACSL. Note that the syntax for comments, continuation, and multiple statements in the language are also followed in the run time commands. The SET commands are used to assign new values to model variables for distribution constants (k_{21} and k_{31}) and to the NH_3 to EO ratio (RATIO). These statements override the values given in the model in CONSTANT statements.

FIGURE 2 - Simulation Commands

```
    'SET VARIABLE VALUES'
SET K21=.5, K31=.25, RATIO=.4

'SPECIFY SOLUTION VALUES TO BE STORED'
PREPARE TIME, MEA, DEA, TEA, EO, NH3

'START A SIMULATION OF MODEL'
START

    'FIGURE 3'
PLOT MEA, DEA, TEA, 'SAME'

    'FIGURE 4'
PLOT EO, 'LOG', NH3, 'SAME'

    'FIGURE 5'
SET TITLE='ETHANOLAMINE REACTION...
 PRODUCT DISTRIBUTION;...
 NH3/EO = .4, AQ HN3 =18%;'
SET XLABEL='TIME (MINUTES)'
SET YLABEL='MEA (BLUE), DEA (RED),...
 TEA (GREEN) - MOL/LITER'
'COLOR AND GRIDDED PLOT'
SET COLCPL=.TRUE., GRDCPL=.TRUE.
PLOT MEA, DEA, TEA, 'SAME'
```

TABLE 5 - Run-Time Commands

Model Data Commands

SET	Sets model variables
DISPLAY	Prints specified variables
*DATA	Specifies experimental data
*DEBUG	Prints current values
SAVE	Saves current values
RESTORE	Restores saved data values
*TERMINAL	Command from terminal
*FILE	Command from disk file

Integration Control Commands

START	Begins integration
CONTINUE	Continues integration
REINIT	Resets initial conditions
MERROR	Sets relative error
XERROR	Sets absolute error

Integration Data Commands

PREPARE	Stores specified values
PLOT	Plots PREPARE data
PRINT	Prints PREPARE data
OUTPUT	Prints specified variables
RANGE	Prints range of PREPARE data

Executive Commands

STOP	Terminates the DACSL session
PROC	Begins a procedure
END	Ends a command procedure
*HELP	Prints information about commands
ACTION	Adds to scheduled actions list
ANALYZE	Calculates and reports steady-state
*TIME	Prints used computer time

Nonlinear Programming Commands

*ESTIMATE	Starts parameter estimation
*RESULT	Prints report at current parameter values
*FIT	Specifies experimental data columns
*VARY	Specifies model parameters
*LOWER	Specifies lower bounds
*UPPER	Specifies upper bounds
*EQUAL	Specifies equality constraints
*MINIMIZE	specifies minimization variable
*MAXIMIZE	Specifies maximization variable

*Commands added to ACSL in DACSL.

The PREPARE statement specifies model variables that are to be stored during the solution the equations for later plotting or printing. The START statement causes the simulation to be performed from the initial independent variable value until the condition specified on the TERMT statement in the model is met. A system variable (IALG) can be assigned a value using the SET command to select an integration algorithm to be used by the START command.

The DISPLAY statement prints the current variable values when entered. A print command is available to print variables from the data specified on the PREPARE command after a simulation. An OUTPUT command will display values during a simulation. A DEBUG command is also available to print current values of all model variables.

The PLOT statement retrieves data specified that was stored during the START command and displays them graphically. The same command syntax can generate line printer plots or pen type plots on a number of different devices like IBM, HP, Tektronix, CalComp and Zeta graphic terminals or plotters. The device is specified in the system command that loads the program for execution. The first PLOT command in the example displays the product concentrations as a function of time (Figure 3). The first variable of the PREPARE command is selected as the independent variable. It can be changed using a PLOT command option. The next plot command shows EO and NH_3 as a function of time on a semi-log plot (Figure 4).

FIGURE 3 - Plotting Example

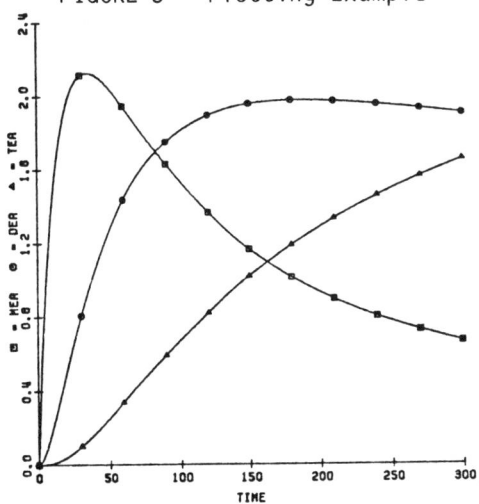

FIGURE 4 - Semi-log Plotting Example

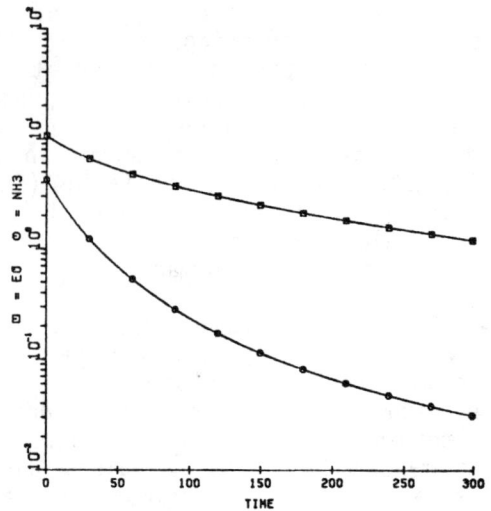

The system contains many system variables that permit a variety of options for graphics such as removal of symbols, lines, axis, plotting a family of curves, and other options. The example in Figure 2 illustrates some of the other options to override the axis labels and include a title, a grid and color on the plot. The system variables are assigned values using the SET command. Figure 5 shows the results of these commands.

FIGURE 5 - Higher Quality Plotting Example

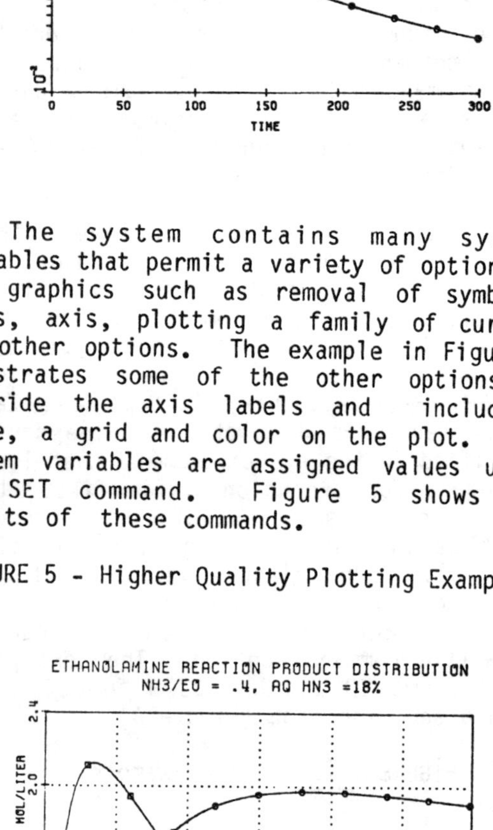

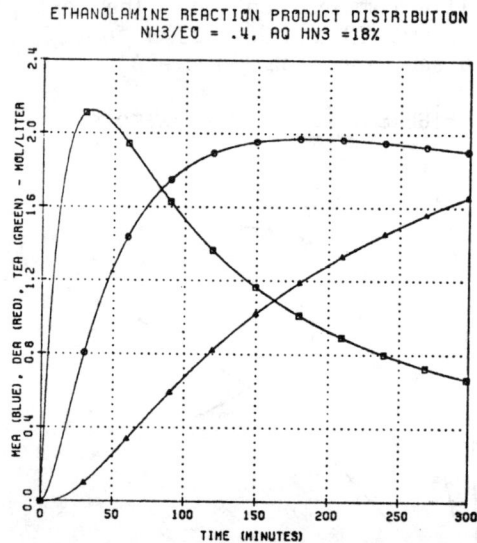

Parameter Estimation Commands. The results shown in the last section were obtained for a model without estimating the parameters from experimental data. Commands for performing this parameter estimation are illustrated in this section and in Figure 6. In it experimental data for distribution of EO in the products are shown as a function of NH_3 and NH_3 to EO ratio. This data set contains many missing data values indicated by a period (.). These values are ignored during calculation of the objective function. The INITIAL keyword in the data command causes the values of variables of integrator initial conditions to be set from the data.

FIGURE 6 - Parameter Estimation Commands

```
DATA
TIME CNVEO MONO  DI   TRI  RATIO NH3AQ TEMP
  0    .     .    .    .    5.1   18   15 INITIAL
 18   37     .    .    .     .     .    .
 60   80     .    .    .     .     .    .
500    .   45.1 32.1 20.1    .     .    .
  0    .     .    .    .   10.1   18   15 INITIAL
 17   31     .    .    .     .     .    .
 18   33     .    .    .     .     .    .
 25   49     .    .    .     .     .    .
500    .   43.4 33.0 20.2    .     .    .
  0    .     .    .    .   13.9   18   15 INITIAL
 25   39     .    .    .     .     .    .
500    .   58.8 28.0 11.0    .     .    .
  0    .     .    .    .   19.7   18   15 INITIAL
 25   43     .    .    .     .     .    .
500    .   70.5   .    .     .     .    .
END

'SPECIFY PARAMETERS TO BE ESTIMATED'
VARY K21, K31
'SPECIFY DATA COLUMNS TO BE FIT'
FIT MONO=0., DI=0., TRI=0.
'PERFORM PARAMETER ESIMATION - FIGURE 7'
ESTIMATE

'RESET TSTOP TO IGNOR LAST POINT'
SET TSTOP=61.
'SPECIFY PARAMETER TO BE ESTIMATED'
VARY K1
'K1 LIMITED TO POSITIVE VALUES'
LOWER K1=0.
'SPECIFY COLUMN OF DATA TO BE FIT'
FIT CNVEO=0.
'PERFORM PARAMETER ESTIMATION - FIGURE 8'
ESTIMATE
```

From this data it is desired to estimate the distribution constants k_{21} and k_{31} by fitting the MONO, DI, and TRI column the distribution of EO in the products MEA, DEA, and TEA respectively. The VARY command is used to specify the estimation variables and FIT to specify the columns of data to be used. The constants in the FIT commands are the heteroscedasticity parameters. The zero values indicate that a least squares objective function should be used. If the values are omitted then optimum values of the heteroscedasticity parameter would be found for each column of data.

The ESTIMATE command in Figure 6 causes the parameter estimation calculations to be performed. The optimization method used is generalized reduced gradient method unless specified otherwise in an ESTIMATE subcommand. Some of the other methods available are Broyden, Davidon-Fletcher-Powell-Shanno, and the complex (constrained Nelder-Mead). Since the objective function is often "noisy" due to integration round-off error, the complex direct search method often works better than the gradient method for problems of four or less parameters. The results of ESTIMATE are shown in Figure 7.

FIGURE 7 - k_{12} and k_{31} Estimation Results

DESCRIPTION	PARAMETER ESTIMATES	
	INITIAL	FINAL
OBJECTIVE FUNCTION	46.15	27.83
K21	1.000	9.411
K31	1.000	11.75

TIME	MONO OBSERVED	MONO PREDICTED	% ERROR	DIFFERENCE	RESIDUAL PLOT
.0		100.0			
18.00		44.91			
60.00		35.75			
500.0	45.10	35.46	21.38	9.643	***
.0		100.0			
17.00		63.93			
18.00		63.06			
25.00		58.60			
500.0	43.40	52.35	-20.61	-8.946	**
.0		100.0			
25.00		66.48			
500.0	58.80	60.16	-2.31	-1.357	*
.0		100.0			
25.00		74.08			
500.0	70.50	68.08	3.43	2.419	**

TIME	DI OBSERVED	DI PREDICTED	% ERROR	DIFFERENCE	RESIDUAL PLOT
.0		.0			
18.00		32.90			
60.00		32.73			
500.0	32.10	32.69	-1.85	-.5937	****
.0		.0			
17.00		27.21			
18.00		27.62			
25.00		29.52			
500.0	33.00	31.58	4.32	1.424	**********
.0		.0			
25.00		25.92			
500.0	28.00	28.91	-3.27	-.9148	*****
.0		.0			
25.00		21.48			
500.0		25.08			

TIME	TRI OBSERVED	TRI PREDICTED	% ERROR	DIFFERENCE	RESIDUAL PLOT
.0		.0			
18.00		22.19			
60.00		31.55			
500.0	20.10	31.85	-58.45	-11.75	**********
.0		.0			
17.00		8.866			
18.00		9.326			
25.00		11.89			
500.0	20.20	16.08	20.41	4.122	*****
.0		.0			
25.00		7.594			
500.0	11.00	10.93	0.65	7.195D-02	*
.0		.0			
25.00		4.445			
500.0		6.840			

FITTING STATISTICS

	MAXIMUM LIKELIHOOD FUNCTION	WT RESID SUM OF SQUARES	WEIGHTED RESIDUAL SUM	STANDARD ERROR OF ESTIMATE	PERCENTAGE VARIATION EXPLAINED	WEIGHTING PARAMETER
OVERALL	-27.83	339.0	-5.881	6.509	89.24	
MONO	-13.30	180.7	1.759	6.722	69.73	.0
DI	-4.362	3.217	-8.450E-02	1.036	80.17	.0
TRI	-10.17	155.0	-7.555	7.189	46.72	.0

The remaining statements in Figure 6 are used to estimate k_1 by curve fitting the percentage EO conversion data (CNVEO). The LOWER command in this section specifies lower limits of a parameter or function of the parameters. Two other commands are also accepted for specifying upper limits (UPPER) and equality constraints (EQUAL). The parameter estimation results are shown in Figure 8.

FIGURE 8 - k_1 Estimation Results

DESCRIPTION	PARAMETER ESTIMATES	
	INITIAL	FINAL
OBJECTIVE FUNCTION	33.80	17.26
K1	5.000D-03	2.175E-03

TIME	CNVEO OBSERVED	CNVEO PREDICTED	% ERROR	DIFFERENCE	RESIDUAL PLOT
.0		.0			
18.00	37.00	34.66	6.33	2.343	*****
60.00	80.00	82.97	-3.71	-2.966	******
500.0		96.37			
.0		.0			
17.00	31.00	31.54	-1.74	-.5379	**
18.00	33.00	33.20	-0.61	-.1999	*
25.00	49.00	44.08	10.05	4.924	*********
500.0		85.29			
.0		.0			
25.00	39.00	43.20	-10.78	-4.204	******
500.0		83.85			
.0		.0			
25.00	43.00	42.46	1.25	.5394	**
500.0		82.45			

FITTING STATISTICS

	MAXIMUM LIKELIHOOD FUNCTION	WT RESID SUM OF SQUARES	WEIGHTED RESIDUAL SUM	STANDARD ERROR OF ESTIMATE	PERCENTAGE VARIATION EXPLAINED	WEIGHTING PARAMETER
CNVEO	-17.26	56.82	-.1019	3.077	97.12	.0

Good initial parameter estimates are essential for convergence to the global maximum of the likelihood function. Graphics are used in DACSL to help find these starting values. When experimental data has been specified by the DATA command, then the symbol on the plot is the experimental data while the line is the predicted model response. By repeated use of the SET, START, and PLOT commands good parameter estimates may be found by manipulating the curve to approximate the experimental data. Figure 9 illustrates this feature. It shows a family of curves comparing the final parameter estimates for the optimization of k_1 and k_{21}, and k_{31} to the percentage EO conversion. (The model shows less variation than the experimental data since water concentration effects were not included in the model.)

FIGURE 9 - Graphical Comparison of Model With Data

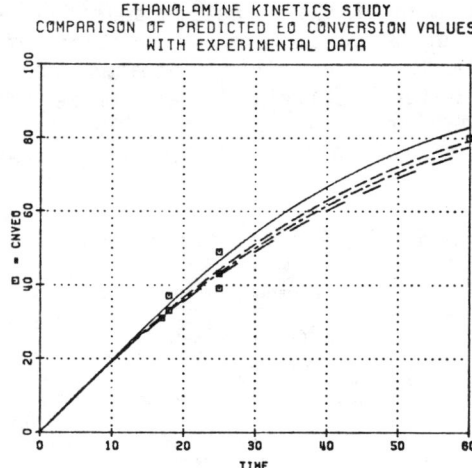

Optimization Commands. In addition to the parameter estimation, parameter optimization can be accomplished within DACSL. The MINIMIZE and MAXIMIZE commands allow the user to specify an objective function directly. These commands replace the FIT parameter estimation command. The LOWER, UPPER, and EQUAL commands can also be used. In an ethanolamine process model that includes water concentration and temperature effects, the yield of a specific reaction product could be optimized through use of these commands.

LITERATURE CITED

1. "ACSL - Advanced Continuous Simulation Language User Guide/Reference Manual", Mitchell and Gauthier Associates, Inc., Concord, Massachusetts, 2nd Edition (1975).

2. Agin, G.L., "Dow Advanced Continuous Simulation Language (DACSL) User Guide/Reference Manual," The Dow Chemical Company (internal use), Midland, Michigan (1981).

3. Bard, Y. "Nonlinear Parameter Estimation," Academic Press, N.Y. (1974).

4. Gear, C.W., "Numerical Initial Value Problems in Ordinary Differential Equations," Prentice-Hall, Englewood Cliffs, N.J. (1971).

5. Kittrell, J.R., "Mathematical Modeling of Chemical Reactors", Advan. Chem. Eng., 8, 97 (1970).

6. Lasdon, L.S., "A Survey of Nonlinear Programming Algorithms and Software," Foundatation of Computer-Aided Chemical Process Design edited by R.S. Mah and W.D. Seider, Engineering Foundation, N.Y., 185 (1980).

7. Lasdon, L.S., A.D. Waren, and M. Rathner, "GRG2 User's Guide," CIS-78-01, Department of Computer and Information Science, Cleveland State University, Cleveland, Ohio (1978).

8. Mitchell, Edward E.L., "Advanced Continuous Simulation Language (ACSL)," Numerical Methods for Differential Equation and Simulation edited by A.W. Bennett and R. Vichnavetsky, IMACS, North-Holland Publishing Company (1978).

9. "Report of the Technical Committee on Continuous System Simulation Languages", Simulation, 9, 281 (Dec. 1967).

10. Miki, M., T. Ito, M. Hatta, and T. Okabe, "Reaction of Ethylene Oxide with Active Hydrogen I.: Reaction of Ethylene Oxide with Ammonia," Yu Kagaku, 15, 215 (1966).

11. Reilly, P.M., R. Bajramovic, G. E. Blau, D. R. Branson and M. W. Sauerhoff, "Guidelines for the Optimal Design of Experiments to Estimate Parameters in First Order Kinetics Models," Can J. Chem. Eng., 55, 614 (1977).

12. Sargent, R.W.H., "A Review of Optimization Methods for Nonlinear Problems," ACS Symposium Series 124 (1979).

REVIEW OF SCHEDULING OF PROCESS OPERATIONS

G.V. REKLAITIS

School of Chemical Engineering
Purdue University
West Lafayette, Indiana 47907

ABSTRACT

The scheduling of multiproduct production facilities is an important problem relevant to a large sector of the processing industry. In this paper an overview is given of the role of scheduling within the overall problem of production planning. Single level and hierarchical planning models are considered. A review is made of the portions of the broad scheduling literature applicable to Chemical Processing Industry problems. Approaches to parallel, serial, and general network process structures are summarized. Finally, implementation strategies employing interactive computer graphics technology are discussed.

INTRODUCTION

In the context of the Chemical Processing Industry (CPI), Production Scheduling is the methodology according to which the order in which products are to be processed in each of the units of a plant is determined so as to optimize some suitable economic or systems performance criterion. Scheduling is always required whenever a processing system is used to produce multiple products by sharing the available production time between products.

As an example, suppose that the multistage production facility of Figure 1 consisting of batch reactors and continuous separators, is used to produce 10 different products each of which requires the same sequence of processing steps but possibly different operating conditions. Since the processing conditions for each product are assumed to be known, the batch residence times and continuous unit processing rates will be fixed. Given that certain quantities of each of the 10 products must be produced over a specified time period, say 30 days, the scheduling problem consists of determining the order in which the products are to be processed in each stage and, thus, the times at which each operation is to be started and stopped so as to minimize the processing cost or, perhaps, to minimize the time required to complete the processing of all products.

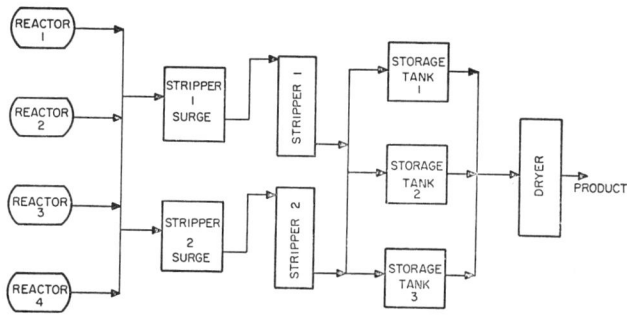

Figure 1. Multi-Stage Production Facility

Note that the need for scheduling does not arise from the nature of the processing operations: continuous, semi-continuous, or batch, nor is it determined by the properties of the processed materials. Scheduling is only required because the available production time on the facility must be subdivided among the various products. The solution of the scheduling problem will be critically affected by the structure of the processing network, the processing times required for each product on each operation, the presence or absence of intermediate storage, the lost production time or cost associated with product change-overs, the cost or performance criterion used to rank schedules, as well as any due dates which are assigned to individual products. At the present time no general solution procedures exist which can accommodate all of the above potential problem features. Instead, an extensive literature exists of solution approaches for specialized forms of the scheduling problem. Most if not all of this literature has its origin in non-chemical processing applications: machine shops, discrete assembly manufacture, computer systems operation, and so on. However, because of the similarities in the abstract problem structure, a considerable segment of this literature is directly applicable to scheduling in the process industries. In this paper we will survey that segment which appears to be relevant to the CPI.

The astute readers will have observed that the form of the scheduling problem described above is rather restrictive (some might say simplistic) because it does not take into account important factors which impact on plant operating policy: raw material or chemical intermediate inventories and production lead times; resource restrictions such as labor availability and maximum allowable utility loads; product inventory control parameters such as reorder positions, safety stock levels, and economic order quantities; equipment maintenance and down times; as well as longer term demand forecasts. Clearly these factors must be coordinated/meshed with the scheduling function in order to arrive at an overall optimum plant production plan. In fact there is a considerable literature in the management science/operations research field which has wrestled with the overall <u>production planning problem</u>. As part of our review of scheduling methodology we will therefore first consider this broader problem in order to elucidate the proper role of scheduling within plant operating policy.

The discussion of this paper will therefore flow as follows: First we will consider the proposed approaches to the overall production planning problem. Next, we will survey scheduling approaches which appear to have relevance to the processing industry. Finally, we will conclude with a brief discussion of approaches for the implementation of scheduling methodology.

PRODUCTION PLANNING

Production planning is the discipline concerned with the allocation of production capacity and time; raw materials, intermediate product and final product inventories; as well as labor and energy resources so as to meet market demands for products over an extended period of time into the future (Hax,1978). Normally for purposes of production planning, it is assumed that the physical resources of the plant are fixed and that the prices, costs, and product demands are externally imposed. Thus, the types and number of production units and inventory facilities, the product slate, the costs of materials, labor, and energy, the product prices as well as the product demands are all assumed to be known or given. The goal of production planning is then to utilize the available resources and plant capacity in the most efficient manner so as to meet the projected longer term demand for products.

The need for production planning exists in all multi-product processing facilities because normally the quantities and frequencies with which product orders are received do not coincide with the production rates at which the plant is most efficiently operated and the ranges of resource levels which are normally available to the plant. In part the role of production planning is thus to "smooth-out" the lumpy demand for products so as to avoid large fluctuations in production rates and resource utilization. The need for production planning is particularly acute when the multiproduct facility must operate under dynamic variations in the product demands. The latter situation is in fact commonly the case with batch/semi-continuously operated plants because it is the flexibility provided by such operations in meeting changing product demands that leads to their selection in the first place. On this basis, quantitative production planning is of interest to the bulk of the processing industry, excluding perhaps only dedicated bulk chemicals plants.

Quantitative approaches to production planning can be classified according to the

number of levels into which the problem is decomposed as well as according to the degree of entity aggregation that is employed. Thus the planning problem can be treated as one single combined problem or as several interrelated subproblems. The products can be aggregated into product families or groups or can all be treated as individual entities. Moreover, planning models can further be categorized according to the assumptions made about by features such as
- i) the structure of the production facility
- ii) the mode of operation of the facility
- iii) the inventory/distribution policies
- iv) the nature of the demand function
- v) the form of the overall production cost function
- vi) the length of the planning horizon

Let us briefly consider several representative examples of these various types of models.

Single Level Approaches

The simplest case of overall planning model is that in which the production facility is treated as one single aggregate unit which produces multiple products over an indefinite horizon with constant deterministic demand. If a cyclic production pattern is assumed and only production set-up and inventory costs are considered, then it is possible to cast the problem as one of selecting the cycle time for each product so as to minimize the cost over a cycle. As surveyed by Elmaghraby (1978), various approximate or partial enumeration schemes have been reported for determining the optimum cycle for each product and the maximum inventory requirements of each product. If the production set-up times are independent of product sequence, then the order in which the products are made can be selected using any convenient secondary criterion. If set-up costs are dependent on the product sequence, then the problem becomes combinatorially complex (equivalent to a traveling salesman problem, Elmaghraby,1978). If the demand is variable over time then a truly cyclic schedule is not possible since the cycle lengths must be readjusted from period to period.

If set-up costs are sequence independent but demand is allowed to vary from time period to time period over a planning horizon consisting of a finite number of time periods, then it is possible to formulate production planning problems solvable by LP methodology (Johnson and Montgomery, 1974). Such models will typically involve three categories of variables:

1. production amount variables: X_{it} = units of product i produced in period t
2. inventory level variables: I_{it} = units of product i in inventory at the end of period t
3. resource level variables: Y_{jt} = units of resource j required in period t

The resource variables may correspond to requirement for manpower, production time, raw materials, or utilities and can be subdivided into normal and extraordinary components (e.g., regular hourly labor and overtime labor). The constraints will include:
1. material balance on each product in each period (production plus change in inventory = demand)
2. balances on the resource utilization of each resource type in each period
3. inequalities involving the available resources and allowable production rates

The objective function will consist of terms corresponding to the variable production costs, inventory holding costs, resource utilization costs, and possibly back-order costs. LP models are obviously attractive because of the flexibility which they offer in accomodating special features of a given application as well as their relative ease of solution. However, the requirements that the objective function be linear in the problem variables and that all variables be continuous do impose important limitations. Moreover, for longer term horizons and applications involving larger numbers of products, the resulting LP's can become very large in size.

To combat these limitations extensions have been reported which include quadratic cost models (especially for work force planning applications, Holt, et al.,1960) as well as general concave (Zangwill,1969) and general nonlinear models(Johnson and Montgomery, 1974). However, the solution of these models is inherently more difficult and the data requirements for their formulation are much higher. The problem of growth in problem size can in part be met by aggregating products into product classes which have similar cost and production requirements. Aggregation has the additional practical benefits of reducing the data requirements for the model, especially the need for detailed multiperiod demand forecasts (Hax, 1978).

Unquestionably the major limitation of

LP planning models is the requirement that all variables be continuous. This requirement disallows treatment of set-up costs as well as direct assignment of products to parallel production facilities. The inclusion of set-up costs immediately requires the definition of 0-1 variables of the general form

$$\delta_{it}(x_{it}) = \begin{cases} 0 & \text{if } x_{it} = 0 \\ 1 & \text{if } x_{it} > 0 \end{cases}$$

Similarly the assignment of parallel processors within each time period requires the addition of further 0-1 variables,

$$Z(i,j,t) = \begin{cases} 1, & \text{if line } j \text{ is used to produce product } i \text{ in period } t \\ 0, & \text{otherwise} \end{cases}$$

While it is easy to formulate models of this type in considerable generality (Pritsker, et al.,1970; Peterson and Silver,1979) the number of zero-one variables which can be routinely accommodated using conventional MIP codes is limited (certainly less than 100 unless the problem is specially structured). When such zero-one variables are imbedded within continuous problems of any realistic size, the computational cost of solution becomes quite unreasonable (Foerster and Reklaitis, 1976). Although in special cases, it is possible to transform the MIP problem to an all integer program solvable via minimal cost network flow methods (Dorsey, et al.,1974), in general, the inclusion of scheduling details (assignment of products to processors) and set-up costs within the overall production planning problem is practical only for small applications.

Hierarchical Production Planning Approaches

A promising alternative approach to carrying out production planning at a single level is to decompose the overall planning problem into a hierarchy of planning/scheduling subproblems (Bitran and Hax,1977). In the simplest form, the problem could be subdivided into an upper level multi-time period planning subproblem such as the LP formulation of the previous section and a lower level sequencing problem in which the specific order in which the products required in each time period are to be produced is determined. The upper level problem would involve production and inventory cost minimization, while the lower level problem might simply involve set-up cost minimization or minimization of the time required to fulfill the production needs for each time-period. The upper level subproblem solution thus serves to define the production quotas and resource utilization levels for each time period, while the lower level solution yields the actual production schedule. The integration of the two subproblems can be carried out in a heuristic fashion (Jain, et al.,1978) but analytical approaches are beginning to receive increased attention (Bitran, et al,1981; Dempster, et al.,1981).

The hierarchical approach is most effectively employed together with a rolling schedule strategy (Hax,1978; Baker,1977). Under this strategy, the multi-time period solution is obtained as before but the scheduling problem is solved and implemented only for the first time-period. One time period later, the multiperiod model is updated and the process is repeated. The rolling schedule strategy thus takes advantage of the look-ahead feature offered by the multi-time period plan which serves to "smooth-out" production requirements. However, it also compensates for the deterministic nature of the planning model since it allows for the continued incorporation of updated information and corrected demand forecasts into the production plan. Analytical and simulation studies reported by Baker (1977) and Baker and Peterson (1979) using simplified planning models indicate that under deterministic demand the solutions obtained with a shorter horizon planning model with the rolling schedule strategy are well within 10% of the optimal cost of the solution obtained from a single long-term planning model. The rolling schedule strategy thus seems to be a practical way of reducing the need for long-term forecasts as well as the expense of solving long-term multi-period problems.

In applications involving a large number of products, it can become desirable to aggregate products into families or classes for purposes of executing the upper level model. If aggregation is used, however, then it may be expedient to introduce additional subproblem levels (Bitran and Hax,1977; Bitran et al.,1981). For instance, once aggregate production levels are determined, a disaggregation subproblem might be solved in which product run lengths might be determined so as to minimize set-up costs. This could then serve to set the constraints for a third level subproblem in which detailed schedules are generated for individual products. Certainly many issues remain to be explored in the design and implementation of hierarchical production planning systems. However, the basic approach is very appealing n that it does mirror the organizational structure within which production plans are implemented: corporate level longer term production planning, development of intermediate term production schedules at the process level and generation of day by day (or hourly) schedules at the operating level. It is reasonably efficient in its use of information about the production system: shorter horizons can be used in the multi-time period model, requiring less demand forecasting, and detailed information is only required for the single time-period subproblems.

Although in the preceding discussion the upper level problem is formulated as an optimization problem, other heuristic or analytical procedures could also be used to set the production quotas for the lower level subproblems. One technique widely used in the manufacture of component assemblies and receiving increasing attention in the processing industry is <u>Materials Requirements Planning</u> (Orlicky, 1975; Miller and Sprague, 1975). MRP is an inventory management and production planning technique which, given a delivery schedule for finished products, determines the initiating times for all raw materials procurement orders, for the production runs needed to prepare all required intermediate products, as well as the starting times for the production of the final product itself. The technique uses three types of process system information:
  i) a tree which indicates what steps and intermediate materials are required to produce each product as well as the material balance factors which specify how much of each raw material or intermediate is required per unit of the final product
  ii) the estimated lead times required to produce or procure each component material from its immediate precursor
  iii) the minimum order quantity to be used for stock replenishment as well as the current inventory level of each component material.

Given a delivery time for a product shipment, each branch of the product processing tree is traced from the product in question and each precursor component requirement is calculated. If it is available in inventory, the inventory of that component is reduced accordingly. If the inventory is inadequate, then a production order is issued for that component at a time calculated using the component lead time. The tracing of each branch of the processing tree continues until all raw materials requirements have either been met or ordered. At this point a time phased sequence of production or raw materials orders will have been assembled for all components required in the production of the desired product. Assuming that the lead times are correctly estimated, the production or stock orders will have been time phased such that all components will be available when they are required at the appropriate point in the production chain for the final product considered. In this way the product delivery date will be met without excess inventory. Of course, the accounting procedure assumes that the delivery schedule imposed can in fact be met with the available production capacity. Hence, the MRP methodology does not constitute a complete production planning model. Nonetheless, the MRP logic does define the due dates and the product requirements for the detailed production scheduling subproblem and thus can, when combined with a delivery date assignment procedure, serve as an upper level planning methodology. As described by Cohen and Zeftel (1980), MRP has successfully been used as a planning tool for the DuPont Chambers Works. Liberatore (1979) discusses an application in FMC's Modesto carbonate production process. Both of these studies suggest that the use of MRP should be linked with detailed process scheduling models and imply that because of the effects of schedule changes on lead times some iteration between the MRP methodology and the detailed schedule generator may be appropriate.

PROCESS SCHEDULING

From the preceding discussion it is evident that scheduling of the actual production runs of the various products on the available processing units is an important component of the overall production planning problem. In general, the scheduling problem will involve the following elements:
  i) a set of N products which must be processed
  ii) a set of M processing units which are available for this purpose
  iii) a performance criterion with respect to which the schedule is optimized
  iv) a matrix of production times (T_{ij}) associated with each product i and unit j
  v) a set of rules which govern the production process including:
    (a) the order in which the operations must be performed for each product
    (b) the manner in which intermediate storage between processing units is regulated
    (c) the allowances which are made for subdividing product runs.

In special cases, the scheduling problem may in addition be formulated with a set of due dates for each product or a matrix of change-over times/costs associated with each ordered product pair. Moreover, in some cases the product requirements may be defined in terms of amounts and the processing capacity in terms of unit rates; but, these are easily converted to processing times. The problem is virtually always assumed to be deterministic, that is, all problem parameters are known in advance, as well as static, that is, the production requirements remain unchanged for the duration of the schedule. New requirements which may arise while the schedule is being implemented are thus saved for the production schedule to be generated in subsequent time periods.

The deterministic and static scheduling problem has associated with it a large and rapidly growing literature. It is certainly impossible to do it justice in a single review paper. We can only call to the readers attention several excellent books and review articles and confine our discussion to the special forms of the problem which are of most interest to the CPI. To that end we recommend the pioneering book by Conway, et al.(1967), the excellent introductory text by Baker (1974), and the more advanced book edited by Coffman (1976). For a fine exposition of the complexity theory which has recently emerged for scheduling problems, the monograph by Gare and Johnson (1979) is highly recommended. A number of review papers on the subject have appeared including the general reviews (Graham et al.,1979; Graves, 1981) and the focused reviews (Baker,1973; Godin,1978; Panwalker and Iskander,1977).

Scheduling problems are most easily classified according to the performance criterion used, by the structure of the processing unit network, and by the specific production rules governing the process. The performance criteria are generally of two types: cost based criteria and system performance based criteria. Cost based criteria are usually only employed with single unit problems since these can often be treated via a single level planning model. Most work with multi-unit processes employs system performance based criteria (Baker,1974; Graves,1981) with the three most commonly employed being:
  i) minimize the total time required to produce all products (often called the makespan)
 ii) minimize the average residence time (called the mean flowtime in the literature) of all products in the system. Sometimes the mean flowtime is calculated using weighting coefficients for the individual products.
iii) minimize the maximum or average tardiness of the schedule, where tardiness is the positive part of the difference between the completion time of a product and its due date.

If the scheduling problem is treated as a subproblem of the production planning problem, then costs will have been taken into account at the upper level and, therefore, system performance based criteria are certainly appropriate. For most applications, the makespan criterion is satisfactory. However, if the production system is overloaded, then an average flowtime or tardiness based criterion will help insure that products are scheduled so as to distribute the delivery delays.

For CPI applications, three types of network structures are of greatest interest:
  i) parallel units
 ii) serial units
iii) generalized serial systems with multiple parallel units per stage.

The very extensively studied single unit problem is of lesser interest in CPI applications because typically processing requires multiple operations. In any case, for this problem solution procedures now are available for all three performance measures including a weighted tardiness criterion (The makespan will of course be fixed for this problem).

Parallel Unit Process

Parallel unit problems occur quite commonly in the CPI. In such applications the parallel units will typically be continuous trains operated semi-continuously (e.g. in polymer processing facilities such as film extrusion (Foerster and Reklaitis,1976; Overturf, et al.,1978). Parallel unit problems can be subdivided according to whether product production runs can be split among units (pre-emption) or whether, once the product run is initiated on a unit, the run must be completed on that unit (non-pre-emption). The non-preemption assumption is clearly appropriate if a minimum number of set-ups must be used. Parallel unit problems are further classified according to the unit properties
  i) identical units
 ii) uniform units
iii) unrelated units

In the uniform case, each unit has associated with it a distinct processing rate which applies to all products so that the processing time of a product is simply its required amount divided by the unit rate. In the unrelated case, the units are dissimilar so that they will have different processing rates for different products. All three cases in both their preemptive and nonpreemptive forms are of interest in CPI applications. However, the identical unit case does occur less frequently and non-preemptive cases are more common than preemptive applications because it is generally desirable to reduce change-overs to the minimum.

As reviewed in (Graham, et al.,1979; Baker, 1973) solution procedures are available for all forms of the identical, preemptable case. For instance, McNaughton (1959) shows that the minimum makespan is given by

$$\mathrm{mac}\ \{\max_i (T_i),\ \frac{1}{M} \sum_i T_i\}$$

and that a schedule can be generated by simply selecting products arbitrarily and assigning them to each unit such that each machine is occupied for the minimum makespan. An example of such a schedule is shown in Figure 2.

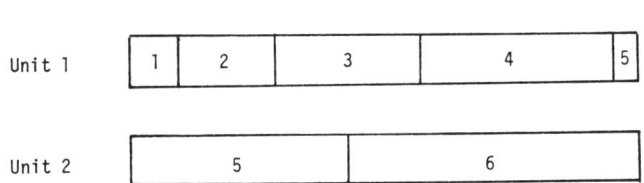

Figure 2. Minimum Makespan via McNaughton's Rule

For the non-preemptable case, the makespan problem can be formulated as a 0-1 integer programming problem (Baker,1973) in terms of variables x_{ij} defined as follows

$$x_{ij} = \begin{cases} 1 & \text{if product i is made in unit j} \\ 0 & \text{otherwise} \end{cases}$$

However, the solution time for this problem grows exponentially with M and N. Problems in which the solution time cannot be bounded by a polynomial in the characteristic size of the problem are said to be NP-complete. The non-preemptable makespan problems with only two identical parallel units is NP-complete (Graham, et al.,1979). For NP-complete problems it is necessary to resort either to approximate methods or else to use enumerative techniques such as branch and bound. In the case of the non-preemptive identical units problem, a good approximate minimum makespan solution can be obtained by sorting the products according to longest processing time first (LPT) and sequentially assigning products to the unit with the currently lowest total processing time. An Example of an LPT schedule is shown in Figure 3.

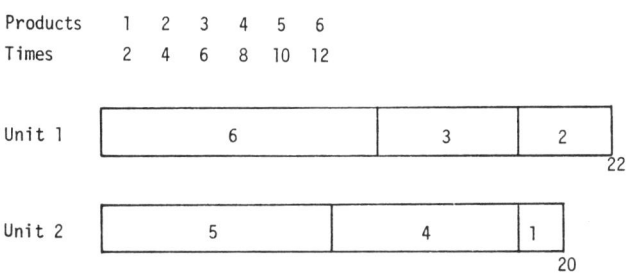

Figure 3. Minimize Makespan for Identical Parallel Units with no Preemption (LPT Heuristic)

A worst case analysis (Graham, et al.,1979) can be used to show that

$$\frac{\text{makespan (LPT)}}{\text{optimum makespan}} \leq \frac{4}{3} - \frac{1}{3M}$$

Interestingly enough, if the mean flowtime criterion is used, then exactly the opposite scheduling algorithm proves optimal. Specifically, the products are sorted according to shortest processing time (SPT) and are assigned sequentially to the unit with the current lowest total processing time (Baker,1974). Moreover, as shown by McNaughton (1959), this algorithm will also be optimal in the preemptive case. An example of a SPT schedule is shown in Figure 4.

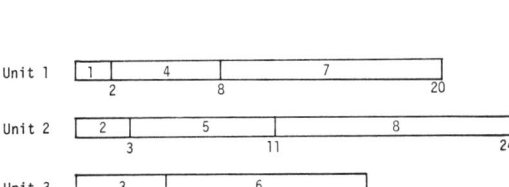

Figure 4. Minimum Mean Flowtime, Identical Parallel Units, using SPT Optimal Rule

The status of algorithms for the twelve forms of the basic parallel unit problem is summarized in Table, based on the compilation reported in (Graham et al.,1974).

Table 1: The Parallel Unit Problem

Types of Units	Makespan		Mean Flowtime	
	PRE-EMPT	NO-PRE	PRE-EMPT	NO-PRE
Identical	McNaughton(1959)	NP-complete	SPT	SPT
Uniform	Gonzales-Sahni (1978)	NP-complete	Gonzalez(1977)	Extended SPT
Unrelated	LP (Lawler & Labetoulle, 1978)	NP-Complete	NP-Complete	Transportation Problem (Bruno, et.al.,1974)

In general, good algorithms exist for all three forms of the preemptive makespan problem and the non-preemptive mean flowtime problem. The unrelated preemptive makespan and non-preemptive mean flowtime problems can be solved as specialized linear programming problems. The NP Complete problems must be solved using approximate or branch and bound techniques. A general dynamic programming solution approach was reported for the non-preemptive case by Rothkopf (1966) but this algorithm is only suitable for small problems. Quite clearly considerable further work is required to obtain good approximate algorithms for the non-preemptive parallel unit case.

Serial Unit Processes

Serial unit processing systems are of course the most prevalent in the CPI. Serial processes involving sequences of batch and semi-continuous units are the most common configurations in multiproduct plants used in borad segments of the food, pharmaceutical, polymer, specialty chemical, as well as paper industries (Loonkar and Robinson,1970; Solem, 1974). Typically, in CPI applications the order in which the processing operations must be executed is fixed by design and is the same for all products because these products tend to belong to a common family (e.g. polyethylene variants). Serial processes with this characteristic are known as flowshops (Baker, 1974). Systems in which each product has a fixed process operation order but this order is different for different products are known as job shops. This class of serial systems occurs to a much lesser extent in the CPI, being restricted largely to small scale or pilot plant operations. Hence, we will confine our discussion to flowshop scheduling problems.

As in the parallel system case, the flowshop scheduling problem can be posed with makespan, mean flowtime, or tardiness based performance criteria. However, most of the reported work has focused on the makespan (Graves,1981). Although both pre-emptive and non-preemptive operation can be envisioned, virtually all of the available results are for the non-premptive case. Finally, because of the staged nature of flowshops, consideration must be given to the status of products between the successive operations. Four types of operating modes are possible:
  a) unlimited intermediate storage between units (UIS)
  b) no intermediate storage between stages (NIS)
  c) zero wait processing (ZW)
  d) finite intermediate storage between units (FIS)

In the unlimited case, intermediate products are assumed to be removed from units as soon as they have completed processing. Unlimited storage is assumed to be available between each processing stage to store such partially completed products. This is the flowshop operation mode which has received the most attention to date. In the no intermediate storage case, intermediate products can be removed from units only if the downstream unit is available for processing. If the downstream unit is engaged, then the intermediate product must wait in the unit in which it has just finished processing. Each unit thus doubles as a storage vessel. This can typically be done with most types of batch processing equipment. In the zero wait case, intermediate products are not allowed to wait between processing steps but instead must be immediately processed in the next unit in the sequence. This operating mode may be appropriate if the intermediate is highly unstable and, thus, must be processed further as soon as it is produced. The fourth case, finite storage, is the same as the first mode (UIS) except that a limit is set on the number of intermediate products which can be stored between processing stages.

An example of a schedule of a three unit-four product flowshop operated under UIS is shown in Figure 5. Note that product 2 must be stored for two time units and product 3 for three times units between units 2 and 3.

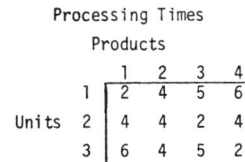

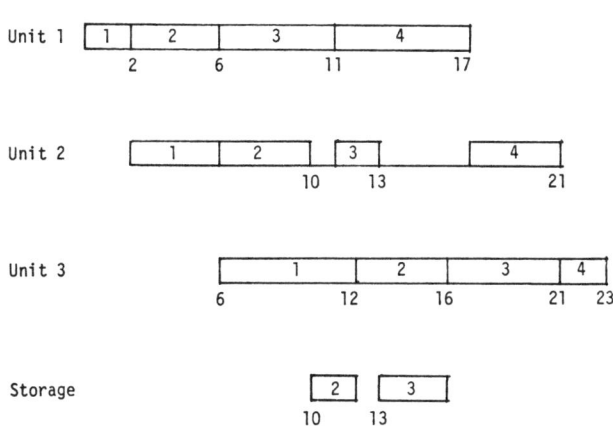

Figure 5. Unlimited Intermediate Storage Schedule

A schedule for the same system operated under NIS is shown in Figure 6a. The shaded segments

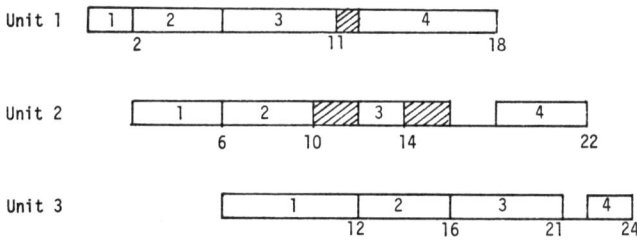

Figure 6a. No Intermediate Storage Schedule

in the bar chart correspond to product being stored in the unit without processing. For instance, at time 10, product 2 has completed processing in unit 2 but must be held for two time increments until unit 3 becomes available. Finally, Figure 6b illustrates ZW operation.

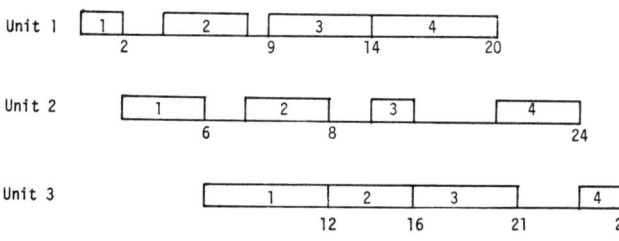

Figure 6b. NIS and ZW Schedules

The characteristic of this mode of operation is, that in order to allow uninterrupted processing of each product, the product start-ups in the first unit must be delayed for a suitable time increment. Note that while the makespan in the UIS case is 23, that of the NIS case is 24 and that of the ZW case is increased to 26. Clearly restrictions on the use of intermediate storage result in increased unit idle times and, thus, increased makespan.

A key feature of flowshop scheduling problems is their apparent simplicity and surprising difficulty of solution. As shown by Garey, et al.(1976), the UIS flowshop problem is NP-complete under the makespan criterion for problems with <u>three</u> or more units and under the mean flowtime criterion for problems with <u>two</u> or more units. The ZW flowshop problem is known to be NP-complete under the makespan criterion for M > 2 and under the mean flowtime criterion for M $\geq$ 2 (Graham, et al.,1979). Suhami (1980) has shown that the NIS flowshop problem is NP-complete for M > 2 . Although, complexity results have not yet been reported for all combinations of performance criterion and operating mode, it can generally be expected that for M > 2 , these problems are NP-complete.

From the preceding summary, it is evident that the only multi-unit flowshop problems for which good solution methods are available are the UIS two stage makespan problem and the ZW two stage makespan problems. The former can be solved using a now classical list scheduling method reported by Johnson (1954). The basis of the schedule logic is <u>Johnson's Rule</u>: In an optimum schedule product i precedes product j if min $\{t_{i1}, t_{j2}\} \leq$ min $\{t_{i2}, t_{j1}\}$. The schedule is generated sequentially as follows. Suppose that the first k products in the sequence and the last ℓ products in the sequence have been identified:

1. find $t_{pq} = \min_{j=1,2} \{t_{ij}$: all i that remain to be selected$\}$
2. If $q = 2$, product p is scheduled $\ell+1^{st}$ from the end
   If $q = 1$, product p is scheduled $k+1^{st}$ in the schedule
3. Product p is deleted from the list of products remaining to be selected and step 1 continues.

An example schedule generated using Johnson's algorithm is shown in Table 2.

Table 2: Makespan for Two Unit Serial System using Johnson's Rule

	Processing Times			
Product	1	2	3	4
Unit 1	2	4	5	6
Unit 2	4	4	1	3

Unscheduled Products	Minimum	Partial Schedule			
1 2 3 4	T_{IJ}	x	x	x	3
1 2 4	T_{32}	1	x	x	3
2 4	T_{11}	1	x	4	3
2	T_{42}	1	2	4	3

An interesting feature of the two stage problem optimal solution is that the order in which the products are processed on the second unit will be the same as that on the first unit. Hence, "passing" of products while in interstage storage will not occur. A similar result is known to be true in the general M unit case for the first two and the last two units in the process. That is, there exists an optimal solution such that the sequence in which the products are processed on the first two units is the same and the sequence in which they are processed in the last two units is the same; however, these two sequences need not be the same (Conway, et al.,1967). As shown by Gupta (1976), the optimal solution of the ZW problem under a generalization of the makespan criterion will have the same product sequence on all units. Even though this result does not hold for flowshop problems in general, it has become conventional in the flowshop scheduling literature to only consider solutions which have the same product sequence on all units. Such schedules are known as _permutation_ schedules. By restricting attention to permutation schedules, the flowshop scheduling problem is reduced to the determination of a single product sequence.

Because of the complexity of flowshop problems, these problems either can be solved exactly using enumerative methods or else can be solved approximately using heuristic methods. Enumerative approaches either involve the branch and bound strategy or else elimination methods (Graham, et al.,1979). Branch and bound strategies typically are based on the use of a partial permutation τ, where τ is a sequence of indices of the products that have already been scheduled. Each branch from the schedule with partial permutation τ is created by adding one product i to that schedule, denoted τi. If $q(\tau,k)$ is the completion time of the partial sequence on unit k, then the completion time of the augmented partial sequence τi will be $q(\tau i,k) = \max\{q(\tau,k), q(\tau i,k-1)\} + t_{ik}$, $1 \leq k \leq M$ where t_{ik} is the processing time of product i on unit k. This recursion can be used to calculate the completion time of the augmented partial sequence τi on the last unit M, $q(\tau i,M)$. Branch and bound strategies differ in how this quantity is used in obtaining a lower bound in the minimum makespan that is possible with any schedule that begins with the product sequence τi (Baker,1975). For instance, one simple lower bound can be obtained by adding to $q(\tau i,M)$ the processing times of the unscheduled products on unit M. However, as summarized in (Baker,1975; Lageweg, et al. 1978) other more elaborate lower bound formulas are available. Since the role of the lower bound is to allow elimination of branches in the enumeration tree, it is desirable to obtain as tight a lower bound as possible. However, this desire must be balanced against the computational time required to evaluate the lower bound.

An alternate enumeration strategy is to use elimination rules to reject possible candidate product sequences. Elimination rules are simply conditions which guarantee that any schedule initiated with one partial permutation will always dominate (have a longer makespan than) a schedule initiated with another partial permutation. One obvious elimination rule is the following (Baker,1975).

If τ_1 and τ_2 are partial sequences containing the same products and if $q(\tau_1,k) \leq q(\tau_2,k)$ for $1 \leq k \leq M$, then τ_1 dominates τ_2.

As a consequence, any schedule beginning with the sequence τ_2 can be eliminated from further consideration. A number of elimination rules have been developed (Szwarc,1971); however, the results of numerical tests conducted by Baker (1975) suggest that branch and bound strategies as a class are superior to elimination methods. More recently, the results reported by Lageweg, et al.(1978) indicate that the use of elimination rules within the branch and bound strategy leads to significant improvement over pure branch and bound enumeration. Nonetheless, the exact enumerative approaches are quite demanding of computer time as the number of units increases. For instance, in the trials reported by Langeweg et al.(1978), fewer than half of the problems involving 20 products and 5 units could be solved within one minute of CPU time on a CDC Cyber 73-28.

An alternative approach to enumeration is to use heuristic techniques to obtain good, suboptimal schedules. A large number of priority rules and heuristic scheduling rules have been proposed for various forms of flowshop and jobshop problems: over 100 are collected and tabulated in (Panwalker and Iskander,1977). One often cited heuristic is that of Campbell, Dudek and Smith (1970) (called the CDS heuristic). The heuristic involves the repeated application of Johnson's two unit algorithm to the general M unit problem. The M unit problem is collapsed to a series of M-1 two unit problems such that the lumped processing times of the kth subproblem are

$$t_{i1}^k = \sum_{j=1}^{k} t_{ij} \quad \text{and} \quad t_{i2}^k = \sum_{j=m-k+1}^{M} t_{ij}$$

The best of the M-1 schedules generated is then selected as the schedule of choice. Computational studies reported by Gupta (1972) suggest that the CDS heuristic is quite good for makespan problems. More recent testing reported by Dannenbring (1977) indicate that a heuristic procedure which uses the following weighting to define a two unit problem solvable using Johnson's algorithm,

$$\theta_{i1} = \sum_{j=1}^{M} (M-j+1)t_{ij}, \quad \theta_{i2} = \sum_{j=1}^{M} (j)t_{ij}$$

followed by a series of pair wise interchanges to improve the solution, is both efficient and relatively insensitive to errors in the processing times. The pairwise interchange method is a depth first search which proceeds as follows. First, generate the N-1 schedules obtained by one pairwise interchange of products adjacent in the sequence. Then, select the best of these and repeat the pairwise interchange. Continue until no further improvement is found. Dannenbring's heuristic was tested on problems involving up to 50 units and 50 products and yielded the optimal solution in 20% of the large problems and gave the best solution out of the 11 heuristics tested in over 70% of the cases.

All of the above methods, exact and approximate, were oriented towards the solution of the UIS form of the flowshop problem. The other forms (NIS,ZW, and FIS) have received very limited attention. The ZW problem was considered by Wisner (1972), Reddi and Ramamoorthy (1972) as well as by Gupta (1976). As shown in these references, the ZW problem can be reformulated to a traveling salesman problem (which itself is an NP-complete problem) and thus can be solved using good enumerative algorithms available for the latter problem. In the case of two unit problems, a special algorithm for the traveling salesman is applicable which is polynomial bounded in solution time (Gilmore and Gombry, 1964). No specialized heuristic procedures have been reported for the ZW problem, although the UIS methods will surely at least give lower bounds to the ZW solution.

The NIS problem was first posed by Levner (1969), who proposed a branch and bound strategy for its solution. However, as noted by Suhami and Mah (1981), the formulation given by Levner may have employed incorrect bounds and, therefore, the enumeration may have been partial and, hence, suboptimal. A probabilistic branch and bound procedure for this problem is reported in Suhami and Mah (1981). The procedure is approximate since lower bounds for partial sequences are established by using probabilistic estimates based upon the Chebyshev inequality. Computational tests on small problems (up to 3 units/10 products) indicate that the strategy is effective in generating schedules within 1% of the minimum makespan.

The FIS problem has received even less attention. A single study involving a two unit flowshop has been reported (Dutta and Cunningham,1975). In the variant studied, completed intermediate products can be either held in intermediate storage or in the unit itself. An exact algorithm based on dynamic programming is presented and because of its computational demands, two approximate solution methods are also outlined. The approximate procedure for "large" problems involves the dynamic programming solution of a series of smaller problems each involving a subset of the jobs. No computational comparisons are reported for any of these methods.

Quite clearly a thorough investigation of algorithms, exact and heuristic would be highly desirable for the ZW, FIS, and NIS forms of the problem. From a practical point of view, it would be interesting to consider forms of the flowshop problem in which some subsequence of units must be operated NW, others NIS, and still others UIS. This would correspond to situations typically arising in batch plants which may involve unstable chemical intermediate products (Egli and Rippin, 1981).

Generalized Serial Systems

As evident from reported design studies of multi-product batch plants (Loonkar and Robinson,1970; Sparrow et al.,1974), such plants are flowshops in which there are multiple parallel units in each serial stage of processing. Such generalized flowshops can be viewed as the natural extension of the parallel unit system and the serial unit system. Unfortunately such systems have received very little attention in the literature: only one study is known to this author (Salvador,1973). To be sure the generalized flowshop problem can be expected to be quite difficult given the complexity of the parallel and series cases. Nonetheless, good approximate solutions are surely obtainable and should be found. The study reported by Salvador (1973) was motivated by the scheduling of a nylon polymerization plant details of which are described in the authors Ph.D. thesis. The problem is posed as one of makespan minimization involving NIS, identical parallel processors in a given stage, permutation schedules only, no preemption, and a strictly increasing sequence of product processing start times. Under these assumptions, the optimum is determined by two sequences: a product sequence and a sequence of product start times. An algorithm is devised that uses a branch and bound search over permutation schedules, which has imbedded within it a dynamic programming solution to determine the product start time sequence. Computational experience is not reported for larger applications but it can be expected that the requirements will be severe. Further work is clearly warranted.

IMPLEMENTATION APPROACHES

From the preceding review of process scheduling methodology it is evident that the solution algorithms available for various forms of the scheduling problem are quite diverse and specific to problem structure. For instance, the identical parallel unit problem under the makespan criterion requires a LPT based schedule while the same problem under the mean flowtime criterion requires a SPT based scheduling approach. In designing a program for routine application in a real production environment one must therefore have the scheduling problem very well defined so that it is very clear which algorithm is appropriate for the process in question. Unfortunately, in the dynamic plant environment the definition of the scheduling problem can change substantially over time. While makespan minimization may be satisfactory when product demands are moderate, tardiness minimization may be required when demands increase. Moreover, specialized constraints, schedule interruptions, and other non-ideal problem constraints may make it difficult to adequately force the resulting scheduling problem to conform to any of the standard problem forms. It thus becomes desirable to have available a library of scheduling methodologies, exact and heuristic, which can be used as required. Moreover, it becomes necessary to structure the programs so that the results of a scheduling run can be analyzed, easily edited, and modified or rerun. The editing feature of such a system is particularly valuable because it permits the possibility of human intervention in the algorithmic process. This can be particularly effective with branch and bound type algorithms in which an experienced user may frequently recognize and eliminate branches that could otherwise require extensive fathoming calculations. These considerations have led to a growing interest in <u>interactive scheduling</u> systems which use computer graphics technology.

Interactive scheduling systems were first considered in the mid-sixties. From the survey prepared by Godin (1978), it is evident that activity in this area has grown considerably in recent years. While much of the work appears to have been oriented towards job shop problems, a flowshop oriented system has recently been reported (McDonald and Hodgson, 1981), as has a prototype system, DURPAK, designed for chemical batch processing applications (Durand,1979). Systems of this type contain many of the following features:
1. convenient problem definition facilities
2. a selection of heuristic and exact algorithms
3. provisions for selective interruption, modification, and restarting of algorithms
4. graphics facilities for viewing schedules and schedule information
5. schedule modification operators for user revision of schedules
6. a rudimentary simulator for simulating the final generated schedule.

While the utility of the first four features are self-evident, the last two bear further elaboration. Schedule modification or improvement operators are useful because they allow an accepted schedule to be revised in an evolutionary manner without major upsets in operations. Moreover, for non-standard situations heuristic local improvement methods may be the only recourse to obtaining good schedules (Godin,1978). As noted by McDonald and Hodgson (1981), a rudimentary schedule simulator is an essential element of such a system since it allows investigation of schedule robustness under potential changes in the production environment and, in general, aids in the visualization of the consequences of the generated schedule. While experience with systems of this type in the CPI is very limited (DURPAK has never been subjected to in plant use), interest in these systems is growing and several large CPI companies are reported to have development projects in progress.

CONCLUSIONS

In this paper we have attempted to show that the production planning problem and its key subproblem, processing schedule generation, constitute an important area to which the focus of chemical engineers should be directed. As shown in this survey a considerable body of techniques are available at this time for quantitative treatment of planning and scheduling problems. For applications typical of the CPI, a hierarchical planning approach which employs an upper level, multi-time period, LP, cost based planning model and one or more lower level single time period scheduling models seems quite appropriate. The hierarchical approach should employ the rolling schedule strategy to accommodate the dynamic market and process environment. Process schedule generation should be accomplished using an interactive scheduling system which incorporates a selection of scheduling algorithms and local improvement techniques. Considerable research remains to be done to develop practical scheduling techniques for generalized multiproduct processing networks and such research is the proper domain of our engineering profession.

ACKNOWLEDGMENT

This work was supported in part under Grant No. CPE-7924565 from the National Science Foundation.

LITERATURE CITED

Baker, K.R. and D.W. Peterson, "An Analytical Framework for Evaluating Rolling Schedules", Manag. Science 25, 341-351 (1979).

Baker, K.R., "A Comperative Study of Flowshop Algorithms", Opns. Res. 23, 62-73 (1975).

Baker, K.R., "An Experimental Study of the Effectiveness of Rolling Schedules in Production Planning", Decision Science 8, 19-27 (1977).

Baker, K.R., Introduction to Sequencing and Scheduling, Wiley, New York (1974).

Baker, K.R., "Procedures for Sequencing Tasks with One Resource Type", Int. J. Prod. Res. 11, 125-133 (1973).

Bitran, G.R., E.A. Haas, and A.C. Hax, "Hierarchical Production Planning: A Single Stage System", Opns. Res. 29, 717-743 (1981).

Bitran, G.R. and A.C. Hax, "On the Design of Hierarchical Production Planning Systems", Decision Science 8, 28-55 (1977).

Bruno, J., E.G. Coffman, Jr., and R. Sethi, "Scheduling Independent Tasks to Reduce Mean Finishing Time", Com. ACM. 17, 382-387 (1974).

Campbell, H.G., R.A. Dudek, and M.L. Smith, "A Heuristic Algorithm for the n Job m Machine Sequencing Problem", Manag. Science, 16, 630-637 (1970).

Coffman, E.G., Jr.(ed), Computer and Job Shop Scheduling, Wiley, New York (1976).

Cohen, R.L. and L. Zeftel, "Materials Requirements Planning in the CPI", CEP 76, 59-63 (1980).

Conway, R.W., W.L. Maxwell, and L.W. Miller, Theory of Scheduling, Addison-Wesley, Reading, Mass. (1967).

Dannenbring, D.G., "An Evaluation of Flowshop Sequencing Heuristics", Manag. Science 23, 1174-1182 (1977).

Dempster, M.A.H., M.L. Fisher, L. Jansen, B.J. Lageweg, J.K. Lenstra, and A.H.G. Rinnooy Kan, "Analytical Evaluation of Hierarchical Planning Systems", Opns. Res. 29, 707-716 (1981).

Dorsey, R.C., T.J. Hodgson, and H.D. Ratliff, "A Production Scheduling Problem with Batch Processing", Opns. Res. 22, 1271-1279 (1974).

Durand, C.X.P., "DURPAK: A Computer Program to Handle the Scheduling of Multi-Product Plants Running Batchwise under Constraints", Ph.D. Thesis, The University of Western Ontario, London, Ontario, Canada (August,1979).

Dutta, S.K. and A.A. Cunningham, "Sequencing Two Machine Flowshops with Finite Intermediate Storage", Manag. Science 21, 989-996 (1975).

Egli, U.M. and D.W.T. Rippin, "Short-Term Scheduling for Multi-Product Batch Chemical Plants", AIChE Meeting, Houston, Texas (April,1981).

Elmaghraby, S.E., "The Economic Lot Scheduling Problem (ELSP): Review and Extensions", Manag. Science 24, 587-598 (1978).

Foerster, C., and G.V. Reklaitis, "Scheduling of Semi-Continuous Parallel Train Processors with Changeover Restrictions", AIChE 69th Annual Meeting, Chicago, IL (1976).

Garey, M.R. and D.S. Johnson, Computers and Intractability, W.H. Freeman and Company (1979).

Garey, M.R., D.S. Johnson and R. Sethi, "The Complexity of Flowshop and Job Shop Scheduling", Math. Opns. Res., 1, 117-129 (1976).

Gilmore, P.C. and R.E. Gomory, "Sequencing a One State-Variable Machine - a Solvable Case of the Traveling Salesman Problem", Opns. Res. 12, 655-679 (1964).

Godin, V.B., "Interactive Scheduling: Historical Survey and State of the Art", AIIE Trans. 10, 331-337 (1978).

Gonzalez, T., "Optimal Mean Finish Time Preemptive Schedules", Technical Report 220, Computer Science Dept. Penn. State University, (1977).

Gonzalez, T. and S. Sahni, "Preemptive Scheduling of Uniform Processor Systems", J. ACM. 25, 92-101 (1978).

Graham, R.L., E.L. Lawler, J.K. Lenstra, and A.H.G. Rinnooy Kan, "Optimization and Approximation in Deterministic Sequencing and Scheduling: A Survey", Ann. Discrete Math. 5, 287-326 (1979).

Graves, S.C., "A Review of Production Scheduling", Opns. Res. 29, 646-675 (1981).

Gupta, J.N.D., "Heuristic Algorithm for Multi-Stage Flowshop Scheduling Problem", AIIE Trans. 4, 11-18 (1972).

Gupta, J.N.D., "Optimal Flowshop Schedules with No Intermediate Storage Space", Naval Res. Logist. Q., 23, 235-243 (1976).

Hax, A.C., "Aggregate Production Planning" in Handbook of Operations Research Models and Applications, J.J. Moder and S.E. Elmaghraby (eds.), Van Nostrand Reinhold Co., New York, N.Y. (1978).

Holt, C.C., F. Modigliani, J.F. Muth, and H.A. Simon, Planning Production, Inventories and Work Force, Prentice-Hall, Eaglewood Cliffs, N.J. (1960).

Jain, S.K., K.L. Scott, and E.G. Vasold, "Orderbook Balancing Using a Combination of Linear Programming and Heuristic Techniques", Interface 9, 55-67 (1978).

Johnson, L.A. and D.C. Montgomery, "Operations Research in Production Planning, Scheduling, and Inventory Control", Wiley, New York (1974).

Johnson, S.M., "Optimal Two-and Three-Stage Production Schedules with Set-up Times included", Naval Res. Logist. Q., 1, 61-68 (1954).

Lageweg, B.J., J.R. Lenstra and A.H.G. Rinnooy Kan, "A General Bounding Scheme for the Permutation Flowshop Problem", Opns. Res. 26, 53-67 (1978).

Lawler, E.L. and J. Labetoulle, "On Preemptive Scheduling of Unrelated Parallel Processors", J. ACM. 25, (1978).

Levner, E.V., "Optimal Planning of Parts Machining on a Number of Machines", Autom. Remote Control, 30, 1972-1978 (1969).

Liberatore, M., "Using MRP and EOQ/Safety Stock for Raw Materials Inventory Control: Discussion and Case Study", Interfaces 9, 1-7 (1979).

Loonkar, Y.R. and J.D. Robinson, "Minimization of Capital Investment for Batch Processes", Ind. Eng. Chem. Proc. Des. Dev., 9, 625-629 (1970).

McDonald, G.W. and T.J. Hodgson, "Interactive Scheduling of a Generalized Flowshop", Industrial and Systems Engr. Dept. Research Report NO.80-16, Univ. of Florida, Gainesville,

Florida, (Oct.1980)(see also the 3-part series in Interfaces, beginning with 11, 42-27 (1981).

McNaughton, R., "Scheduling with Deadlines and Loss Functions", Manag. Science 6, 1-12 (1959).

Miller, J.G. and L.G. Sprague, "Behind the Growth in Materials Requirements Planning", Harvard Business Review, 53, 83-91 (1975).

Orlicky, J., Material Requirements Planning, McGraw-Hill, New York (1975).

Overturf, B.W., G.V. Reklaitis, and J.M. Woods, "GASP IV and the Similation of Batch/Semi Continuous Operations: Parallel Train Process", Ind. Eng. Chem. Process Des. Dev., 17, 166-175 (1978).

Panwalker, S.S. and W. Iskander, "A Survey of Scheduling Rules", Opns. Res. 25, 45-61 (1977).

Peterson, R. and E.A. Silver, Decision Systems for Inventory Management and Production Planning, Wiley, New York (1979).

Pritsker, A.A.B., L.J. Watters, and P.M. Wolfe, "Multi-Project Scheduling with Limited Resources: A Zero-One Programming Approach", Manag. Science 16, 93-108 (1970).

Reddi, S.S. and C.V. Ramamoorthy, "On the Optimal Flowshop Sequencing Problem with No Wait in Process", Opns. Res. Q. 23, 323-331 (1972).

Rothkopf, M.H., "Scheduling Independent Tasks on Parallel Processors", Manag. Science 12, 437-447 (1966).

Salvador, M.S., "A Solution to a Special Class of Flowshop Scheduling Problems", Proc. Symposium in Theory of Scheduling and Application, Springer Verlag, Berlin, 83-91 (1973).

Solem, O., "Contribution to the Solution of Sequencing Problems in Process Industry", Int. J. Prod. Res., 12, 55-75 (1974).

Sparrow, R.E., G. Forder, and D.W.T. Rippin, "Multi-Batch: A Computer Package for the Design of Multi-Product Batch Plants", The Chemical Engineer, 289, 520-525 (Sept.1974).

Suhami, I., Ph.D. Thesis, Northwestern University, Evanston, IL (1980).

Suhami, I. and R.S.H. Mah, "An Implicit Enumeration Scheme for the Flowshop Problem with No Intermediate Storage", AIChE Meeting Houston, Texas (April,1981).

Szwarc, W., "Elimination Methods in the mxn Sequencing Problem", Naval Res. Logist. Q., 18, 295-305 (1971).

Wisner, D.A., "A Solution of the Flowshop Scheduling Problem with No Intermediate Queues", Opns. Res. 20, 689-697 (1972).

Zangwill, W.I., "A Backlogging Model and Multi-Echelon Model of a Dynamic Economic Lot Size Production System - a Network Approach", Manag. Science 15, 506-527 (1969).

PROSIT — AN INTERACTIVE PROCESS SCHEDULING SYSTEM

J.M. NEVILLE
R. VENTKER
T.E. BAKER

Exxon Corporation
Communications & Computer Sciences Department
180 Park Avenue
Florham Park, New Jersey 07932

As described in the literature, (Godin, 1978), operational interactive scheduling systems were quite scarce prior to 1978. Since that time, Exxon Corporation's Applied Math Group has been involved in a project to develop interactive systems which deliver heuristic scheduling algorithms to various scheduling functions. During the project, several prototype systems have been produced as well as a number of unique heuristic scheduling algorithms.

This paper describes the currently operational process scheduling system. After a program of system development and feasibility testing, the new system, called PROSIT (PROcess Scheduling with Interactive Technology), has recently been released for use to the worldwide Exxon circuit. A description of how the system is used by a chemical engineer/process scheduler is provided, including an example, and the paper concludes with a description of application experiences to date.

PROSIT SYSTEM OVERVIEW

The PROSIT system provides a process scheduler with a powerful, convenient tool to aid in activities such as developing feasible schedules, optimizing schedules, making changes, or asking "what if" questions. The system automates most of the calculations currently performed by hand and frees the scheduler to consider many more scheduling alternatives before making decisions. In addition, PROSIT allows the scheduler to make use of advanced automatic scheduling algorithms for developing complex schedules or optimizing existing feasible schedules.

PROSIT essentially provides a computer assisted environment to support the process scheduler in his activities. Schedules can be developed and changed interactively with the impact and result of a change being shown in a matter of seconds. The user sets his own constraints and the system is flexible enough to react to most of his business driving forces.

The PROSIT user interactively describes his problem to the system in terms of units, operations, grades, and storage facilities. Fundamental data such as capacities, operation recipes, supplies, and demands are also required. Optionally, the scheduler can consider complex storage allocation schemes, variable feed rates, composition dependent yields, losses, and various costs such as storage, operation, switching, down time, or runout/overflow penalties. Schedule development is then performed interactively on a spatial representation scheduling screen which immediately shows the impact of a proposed schedule in terms of runouts, overflows, availabilities and cost over the scheduling horizon. This screen also reports the improvement due to using the scheduling algorithms. Finally, there are a number of user oriented features such as a HELP command and a case bank for saving schedules.

Figure 1 illustrates the user's interaction with PROSIT, which is broken into four basic elements: data, solution, simulation, and algorithms. Through a CRT, the user manipulates either the data (which defines the problem) or the solution (the current schedule). The deterministic simulator is invoked to evaluate the new information and produce a new solution. The algorithms, selectively invoked by the user, may operate on an approximate model but, as indicated by the TEST loop, solutions are checked through the simulator. In general, the user is given complete control over schedule development but can relinquish selected tasks to the algorithms as he sees fit.

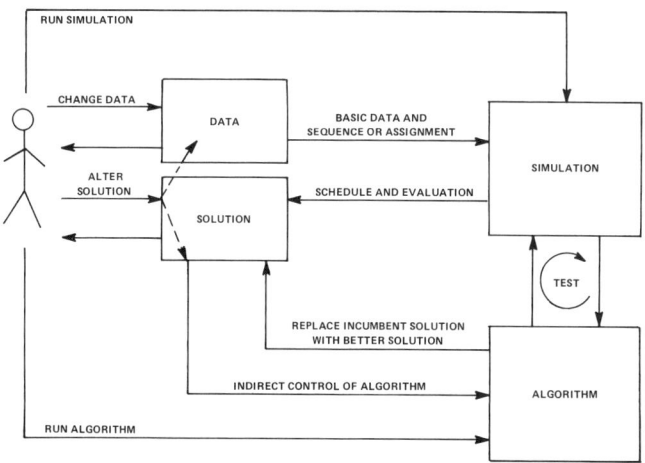

USE OF PROSIT

Use of the PROSIT system is described here in the way a typical problem would be addressed. First, the user must describe his problem to PROSIT through a series of data commands. When the data are complete, the scheduler proceeds to the main scheduling screen and begins building a schedule. He may either use a trial-and-error approach, relying on his own understanding of the problems, or he may invoke one of the scheduling algorithms. Frequently, users will build a nearly feasible schedule and then use selected algorithms to tune it.

In this section, a description of the basic data commands is given in general terms. This is followed by a description of the algorithms and finally an example of the main scheduling screen. The next section provides an example of system use.

Data Commands

Execution of each data command results in the display of a screen. Most are in the form of data tables where the row and column headings depend on an earlier description of the problem. The user then simply fills in the necessary data and moves to the next command. The following discussion is organized in the way a typical problem might be defined.

The DATE command is used first. Here, the user provides the beginning and ending dates for the scheduling period and the demand period. These values provide table and display headings for the time dependent data tables. Later, the user can return to this screen to roll the schedule forward in time and work into the future from existing data.

The DEFINE command is next, and is used to provide mnemonic codes for all entities in the user's problem. These include processing units, possible operations (blending recipes, process yields, etc.), storage facilities (tanks, warehouses, etc.), and grades (raw materials, intermediate and finished products). All other data tables and displays are driven by the entries in this command.

Demands can be entered with the DRATE command. Constant demands (negative) or supplies (positive) are specified and remain in effect throughout the demand period. Optionally, the user can terminate or change a demand at up to ten user specified intervals. Here, the column headings depend on the DATE command while the row headings are the grades specified in the DEFINE command.

Next, the STOR command is used to describe grade-storage facility allocation options. This table is an incidence table where the columns are facilities and the rows are grades. The user inserts codes at the appropriate intersections. Options are:

o Grade is allocated to capacity

o Grade may be allocated to capacity

- Determined as needed by system
- Specified later by user

o Grade is included in dummy capacity to follow combined total of several discrete physical facilities

o Grade is a component in a facility containing a mixture

o Grade is a process yield whose rate depends on the composition in a facility

Storage facilities are further described with the STIN command. Starting inventory and ending inventory targets are provided along with minimum and maximum capacity. Storage cost, overflow penalty and runout penalty can also be specified.

Operation recipes are supplied with the OPER command. Entries here describe the rate and source of raw materials consumed along with the rate and destination of products produced for each operation listed in the DEFINE table. Optionally, the user can indicate that the process yield structure will depend on feed composition and will be calculated by the simulator. It is important to note that this command defines the plant network, process efficiencies, processing capacities, and process capabilities, all at once. In this table, the column headings are operations from DEFINE while the row headings are possible grade-capacity combinations described in STOR.

Operations are further defined with the OPIN command. Here, minimum run length, maximum run length and average run length are specified. Also, optionally, an operation cost can be given in currency per unit time. These data are supplied for each operation in the DEFINE table.

The final command required for basic problem definition is called UNIT. Here, the user describes which operations may be run on which units. Column headings are units while rows are operations, and an "X" at the intersection indicates an allowable combination.

All data commands are summarized in Table 1, and optional commands are described in more detail in the Appendix. As one can see, decomposing problem descriptions into data tables provides interactive problem definition as well as a modular structure for adding new capabilities.

TABLE 1

COMMAND	FUNCTION
DATE	Demand and scheduling periods
DEFINE	Problem entities
DRATE	Supply and demand (constant)
OPER	Operation recipes
OPIN	Operation data
STIN	Storage facility data
STOR	Storage allocations
UNIT	Unit-Operation incidence
BLEND	Blend compositions
BULK	Bulk supply/demand
CSTOR	Combined capacities
DICT	Mnemonic decoding
DOWN	Shutdown costs
LINCOM	Operation yields as linear combinations of feeds
OFFSPEC	Offspec production from switching
OPCO	Operation costs
OSCALE	Process scaling
STPR	Storage preferences
SWCOST	Switching costs
SWLOSS	Switching losses
TARGET	Inventory targets
USCALE	Unit scaling
YIELD	Operation yields from blends

Algorithms

There are six scheduling algorithms currently available in PROSIT. They may all be used to manipulate an existing schedule to reduce cost. Some may also be used to build a schedule with no previous proposal. Each is described briefly, below, and Table 2 lists them all.

TABLE 2

ALGORITHM	OBJECTIVE
DYP	Minimize switching cost for single unit problem
TAP	Vary process run length to minimize cost
TIP	Optimal distribution of run lengths for constant total
ZING	Solve inventory feasibilities by operation insertion and run length manipulation
ZIP	Vary process sequence and run lengths to minimize cost
ZAP	General network code, branch and bound search for optimal solution

For single unit, multi-product problems, a new schedule can be generated by calling DYP (Jones, 1980). DYP is a dynamic programming algorithm which operates on switching costs and uses inventory feasibility constraints to limit the state space. DYP is especially efficient on highly constrained problems.

An algorithm called TAP is available to manipulate run length in order to reduce cost. TAP considers the current sequence of operations as given and treats the run lengths of scheduled operations as independent variables. TAP employs a nonlinear programming algorithm which first determines a gradient direction by finite difference and then performs a line search in that direction before returning control to the user.

A similar algorithm called TIP can be used if the total run length must not be changed. The result of this algorithm is an optimal distribution of the existing operations for the overall run length.

ZING is an algorithm which tries to solve inventory problems by inserting operations. Once a new sequence is determined, the TAP algorithm is automatically used to adjust the run lengths. It is possible, with this algorithm, to focus on a single storage facility, whose infeasibility should be solved.

A fifth possibility is to use an algorithm called ZIP which reorders the current sequence to reduce total cost. No operations are inserted or deleted. ZIP treats the sequencing problem as a traveling salesman problem in which each unit is a salesman and the operations assigned to the units are the cities visited by each salesman. A variation of Lin's method is used. Since the current run lengths may not be valid for a new sequence ZIP automatically uses TAP to adjust the run lengths for each trial sequence.

Finally, a generalized network code is available in an algorithm called ZAP. ZAP attempts to generate a complete new schedule by simultaneously allocating operations for all units for the entire period. A branch and bound algorithm is used to determine which potential solutions to evaluate. This algorithm has been described in detail (Baker, 1981).

These algorithms offer a variety of choices for the scheduler. He can retain control over much of the schedule and focus the mathematics on a specific area. Naturally, if he so desires, he may invoke the overall optimization algorithm.

Scheduling Screen

The main scheduling screen is shown in Figure 2 and is used for schedule development. It provides a spatial representation of the schedule and thereby presents the user with much more valuable data in visual form than could be provided in table form.

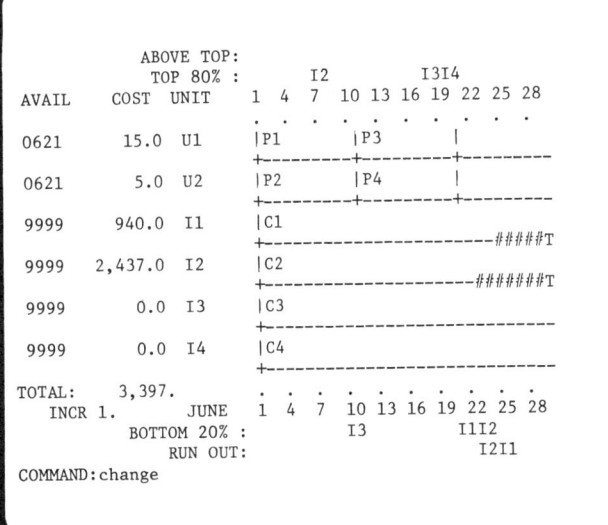

Reading down the screen, line by line, the following is shown:

o No storage facilities have been filled above maximum capacity

o Three facilities have reached the top 80% of capacity

 - I2 on day 7
 - I3 on day 18
 - I4 on day 20

o Unit U1 is

 - Available for additional allocations on day 21
 - Costing $15,000
 - Running operation P1 from day 1 to day 11
 - Running operation P3 from day 11 to day 21

o Unit U2 is

 - Available on day 21
 - Costing $5,000

- Running P2 from day 1 to 11
- Running P4 from day 11 to 21

o Capacity I1 is

- Not available for additional allocation before the end of the scheduling period
- Costing $940,000
- Contains grade C1
- Is infeasible from day 25 on

o Capacity I2 is

- Not available
- Costing $2,437,000
- Contains C2
- Is infeasible starting day 23

o Capacity I3 is

- Not available
- Costing $0
- Contains C3

o Capacity I4 is

- Not available
- Costing $0
- Contains C4

o Total cost is $3,397,000

o An increment of 1 column per day is shown

o Three facilities fall below the bottom 20% of capacity

- I3 on day 10
- I1 on day 21
- I2 on day 22

o Two capacities run out

- I2 on day 22
- I1 on day 25

The existing schedule can be changed simply by moving characters about the allocation lines on the screen. In this case, one could simply allocate P1 to U1 and P2 to U2 both on day 22. Alternatively one could increase the initial run lengths, although there is a risk of overflowing tank I2. A complete example will be presented later.

The Simulator

Entering the CHANGE command to refresh the main scheduling screen executes the Simulator to determine the impact of the proposed schedule. The Simulator proceeds through the scheduling time period in steps and, using the information supplied in the data commands, calculates the result of the proposed schedule in terms of inventories and costs.

The Simulator is written in a very general way and functions in a "data driven" manner. That is, the user literally tells it how to simulate his processing facilities by the information supplied in the data commands.

Support Commands

Finally, there are a number of user support commands. They are listed in Table 3.

TABLE 3

COMMAND	FUNCTION
CPUTIME	Logs computer time and cost
DEBUG	Traces calculations
HELP	Provides command narratives
PROJECT	Reports inventory levels in facilities through scheduling period
STOCKS	Reports grade inventories
?	Further description of error message
paging	Variety of commands to move the screen window through a large table
BRIDGE	Modify case banks across releases
MASTER	Case bank index
PROTECT	Case password protection
RESTORE	Restore a case
SAVE	Save a case
SYSTEM	System control parameters
	— Inventory rule
	— Simulation increment
	— Data Units
RUNPLAN	Specialized scheduling screen

EXAMPLE

A simple example will show how PROSIT is used. Problem decomposition into the data tables is illustrated, as well as schedule manipulation with the main scheduling screen. The following text has been extracted from the current user's manual.

Pipeline Scheduling

Assume that a refinery is supplied with four crudes via two small pipelines. Each crude is stored in a dedicated tank at the refinery. Our problem is to schedule for the month of June the pumping of crude through these pipelines to the refinery so that it does not run out of crude.

Assumptions

Assume that each pipeline can supply all four crudes.

In cases of runouts, a lost profit occurs. Spillovers in the tankfarm, however, are to be avoided under all circumstances.

Transit time is neglected.

Data

The storage tanks have a capacity of 100 KBBl.

Daily consumption of the four crudes is:

- o crude 1: 5 KBBl,
- o crude 2: 7 KBBl,
- o crude 3: 2 KBBl,
- o crude 4: 4 KBBl.

At the beginning of the scheduling period -- now -- the tanks have the following inventory:

- o tank for crude 1: 20 KBBl,
- o tank for crude 2: 60 KBBl,
- o tank for crude 3: 40 KBBl,
- o tank for crude 4: 70 KBBl.

Runout penalties are defined as the lost profit from lack of crude:

- o crude 1: 2 $/BBl/Day,
- o crude 2: 3 $/BBl/Day,
- o crude 3: 4 $/BBl/Day,
- o crude 4: 5 $/BBl/Day.

Both pipelines have a throughput capacity of 10 KBBl/Day.

Experience has shown that a crude batch through a pipeline should be a minimum of three days. From the past operation the following average run lengths have been observed:

- o crude 1: 5 days,
- o crude 2: 8 days,
- o crude 3: 5 days,
- o crude 4: 4 days.

When switching from crude to crude, switching costs occur as follows:

o From crude 1 to

- crude 2: 10 K$
- crude 3: 15 K$
- crude 4: 20 K$

o From crude 2 to

- crude 1: 10 K$
- crude 3: 5 K$
- crude 4: 5 K$

o From crude 3 to

- crude 1: 1 K$
- crude 2: 1 K$
- crude 4: 5 K$

o From crude 4 to

- crude 1: 1 K$
- crude 2: 1 K$
- crude 3: 1 K$

If the pipeline was not used a setup charge of 5 K$ is incurred.

Problem Formulation

There are four crudes, say: C1, C2, C3, and C4, to be scheduled. The two pipelines can be considered as units U1 and U2. An operation is the movement of crude through the pipeline. In this problem there are four operations, say: P1, P2, P3, and P4. Each operation is characterized as a movement of a crude to the refinery into its dedicated tank. (Compare to Figure 3.)

The amount of crude received per day is given for each operation as 10 KBBl (pipeline capacity). The storage tanks have maximum capacities of 100 KBBl. Runout costs and minimum run lengths are defined; opening inventories are specified.

The problem is pictured in Figure 3.

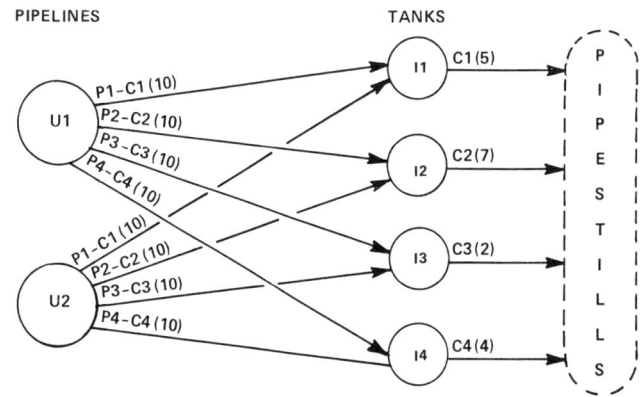

Incidence Tables

DEFine. With the DEFIne screen we specify the units (pipelines), operations (crude batches), storage (crude tanks at the refinery), and the grades (crudes) for this problem:

UNITS (RESOURCES)	OPERATIONS (ACTIVITIES)	STORAGE (CAPACITIES)	GRADES (INVENT.)
U1	P1	I1	C1
U2	P2	I2	C2
	P3	I3	C3
	P4	I4	C4
	XX		

Note that the idle operation XX is also defined. If XX is scheduled, no crude is transported.

STOR. The STOR screen allocates the crudes to the four tanks:

GRADES	CAPACITY			
	I1	I2	I3	I4
C1	S			
C2		S		
C3			S	
C4				S

Note that the body entries are 'S' and not 'X'. This is to signal that the grade/capacity allocations are in effect at the beginning of the scheduling period. If 'X' is used, a manual allocation of grades to capacities becomes necessary.

OPER. This table defines all operations we can use in scheduling the pipeline. When calling OPER the rim of the table is already defined and we enter '10' where appropriate:

CAP-GRAD		OPERATN			
		P1	P2	P3	P4	XX
I1	C1	10				
I2	C2		10			
I3	C3			10		
I4	C4				10	

UNIT. In this problem all operations can be performed on all units. This is the default situation assumed as soon as DEFine is processed. Thus, it is not necessary to define this table. (Of course you can take a peek and display it on the screen.) It looks like this:

OPERATN	UNITS	
	U1	U2
P1	X	X
P2	X	X
P3	X	X
P4	X	X
XX	X	X

Data Tables

The OPER table serves as incidence and data table. It defines the inventory change per day for each operation. The other data given in the business problem are entered in the following tables:

DATE. DATE defines the demand and scheduling period. This problem is formulated without any fluctuations over time, so that the updating of the DATE screen serves only the purpose of using correct dates for table row/column labels:

```
METRIC OR AMERICAN (M OR A):    A
     BEGINNING DEMAND PERIOD:   0601
        ENDING DEMAND PERIOD:   0630
 BEGINNING SCHEDULING PERIOD:   0601
    ENDING SCHEDULING PERIOD:   0630
```

DRATe. Crude consumption in distillation is given for each crude. These data are entered into DRATe as:

GRADES	FLOWRATE
C1	-5
C2	-7
C3	-2
C4	-4

OPIN. Operational data is defined in this array. The problem mentions a minimum run length of three days and specifies average run lengths used in the past. To allow the use of ZAP we need to define maximum run lengths for the batches. Let's assume 20 days. We also increase the average run length to 10 to favor longer crude runs:

OPERATN	MIN RUN (D)	MAX RUN (D)	...	AVG RUN (D)
P1	3.0	20.0		10.0
P2	3.0	20.0		10.0
P3	3.0	20.0		10.0
P4	3.0	20.0		10.0
XX	.	20.0		10.0

STIN. Table STIN holds data related to the storage of products (crudes). The opening inventory, tank capacities, and runout penalties are entered here.

Model data have to be in consistent dimensions. So far we assumed implicitly that the crude volume unit is KBB1, 1000 Barrels, and that the demand and operational data are in KBB1/Day. The runout penalties were defined as $/BB1/Day whereas the STIN screen requires the cost data to be in $/T/Y, where T represents the activity dimension which is KBB1. We need to convert the given cost by multiplying with 365000. This gives numbers which are too large, so let us use K$ (1000 $) as the monetary units. Thus we get the following runout penalties:

- Crude 1:   730 K$/KBB1/Year,
- Crude 2:   1095 K$/KBB1/Year,
- Crude 3:   1460 K$/KBB1/Year,
- Crude 4:   1825 K$/KBB1/Year.

With this adjustment we are now ready to fill out STIN:

CAPACITY	STAR INV (T)	...	CAP MAX (T)	...	LOW PEN ($/T/Y)
I1	20.000		100.000		730.0
I2	60.000		100.000		1095.0
I3	40.000		100.000		1460.0
I4	70.000		100.000		1825.0

The other columns of STIN remain at ' . '. Note that we can change the symbols used to define the dimension on the SYSTem screen. Let's just assume that $ stands for 1000 US dollars and T means 1000 Barrels.

SWCOst. Cost for switching from crude to crude in scheduling the pipeline are entered in this array. Note that the currency unit is K$ and that data in the SWCO array are in K currency units, i.e., in million $:

```
FROM
OPR   P1        P2       P3       P4       XX
P1    1,000.0   0.01     0.015    0.02     .
P2    0.01      1,000.0  0.005    0.005    .
P3    0.001     0.001    1,000.0  0.005    .
P4    0.001     0.001    0.001    1,000.0  .
XX    0.005     0.005    0.005    0.005    1,000.0
```

Note that by default a switch to itself is costing 1,000 K$ to exclude this possibility for ZAP.

Scheduling Screen

By entering the MAIN or CHANge command, the MAIN scheduling screen is displayed as shown in Figure 4.

```
           ABOVE TOP:
             TOP 80% :
AVAIL    COST     UNIT   1  4  7  10 13 16 19 22 25 28
0601     0.0      U1     |.  .  .  .  .  .  .  .  .  .
                         +---------------------------
0601     0.0      U2     |
                         +---------------------------
9999     6,673.3  I1     |C1
                         +---##################T
9999    10,524.0  I2     |C2
                         +-------##################T
9999     1,413.3  I3     |C3
                         +-----------------########T
9999     4,731.6  I4     |C4
                         +---------------########T
TOTAL:  23,342.          .  .  .  .  .  .  .  .  .  .
    INCR 1   JUNE        1  4  7  10 13 16 19 22 25 28
         BOTTOM 20% :    I1    I2    I3  I4
              RUN OUT:   I1    I2        I4 I3
COMMAND:change
```

It shows the inventory situation when no crude is supplied: Slowly the tanks are running out, resulting in huge scheduling cost due to run out penalties. Note that I1-I4 are automatically allocated to crudes C1 to C4 because we entered 'S' in STOR.

Let us assign P1 and P3 to unit U1, and P2 and P4 to unit U2 as shown in Figure 5.

```
           ABOVE TOP:
             TOP 80% :
AVAIL    COST     UNIT   1  4  7  10 13 16 19 22 25 28
0601     0.0      U1     |P1 P3 .  .  .  .  .  .  .  .
                         +---------------------------
0601     0.0      U2     |P2 P4
                         +---------------------------
9999     6,673.3  I1     |C1
                         +--##################T
9999    10,524.0  I2     |C2
                         +-------##################T
9999     1,413.3  I3     |C3
                         +-----------------########T
9999     4,731.6  I4     |C4
                         +---------------########T
TOTAL:  23,342.          .  .  .  .  .  .  .  .  .  .
    INCR 1   JUNE        1  4  7  10 13 16 19 22 25 28
         BOTTOM 20% :    I1    I2    I3  I4
              RUN OUT:   I1    I2        I4 I3
COMMAND:change
```

When hitting enter, the next screen shows the result of scheduling the operations in that fashion for the average run length of 10 days. This is shown in Figure 6.

```
              ABOVE TOP:
                TOP 80% :       I2         I3I4
AVAIL    COST   UNIT    1  4  7 10 13 16 19 22 25 28
                        .  .  .  .  .  .  .  .  .  .
0621     15.0   U1     |P1       |P3         |
                       +---------+-----------+------
0621      5.0   U2     |P2       |P4         |
                       +---------+-----------+------
9999    940.0   I1     |C1
                       +-----------------------#####T
9999  2,437.0   I2     |C2
                       +-------------------------######T
9999      0.0   I3     |C3
                       +-----------------------------
9999      0.0   I4     |C4
                       +-----------------------------
TOTAL:  3,397.
   INCR 1.      JUNE    1  4  7 10 13 16 19 22 25 28
             BOTTOM 20% :              I3    I1I2
                RUN OUT:                     I2I1
COMMAND: change
```

The run outs have been considerably reduced. I2 is filled to the top 80% on day 7; I3 is filled to the top on day 20. We extend the run lengths of the processes as indicated to see whether we hit tank tops. Figure 7 shows how this is done.

```
              ABOVE TOP:
                TOP 80% :       I2         I1        I3
AVAIL    COST   UNIT    1  4  7 10 13 16 19 22 25 28
                        .  .  .  .  .  .  .  .  .  .
0625     15.0   U1     |P1       |P3         |
                       +---------+-----------+------
0623      5.0   U2     |P2       |P4         |
                       +---------+-----------+------
9999      0.0   I1     |C1
                       +-----------------------------
9999    847.0   I2     |C2
                       +--------------------------##T
9999      0.0   I3     |C3
                       +-----------------------------
9999      0.0   I4     |C4
                       +-----------------------------
TOTAL:    867.
   INCR 1.      JUNE    1  4  7 10 13 16 19 22 25 28
             BOTTOM 20% :           I3 I4       I2 I1
                RUN OUT:                           I2
COMMAND: change
```

We are lucky: Inventories remain feasible. Indeed, by pushing out P3 by a few days, we alleviated the I3 inventory problem. To avoid run outs at the end of the scheduling period, we allocate again P1 to U1 and P2 to U2 as shown in Figure 9.

```
              ABOVE TOP:
                TOP 80% :       I2         I3I4
AVAIL    COST   UNIT    1  4  7 10 13 16 19 22 25 28
                        .  .  .  .  .  .  .  .  .  .
0621     15.0   U1     |P1       |P3          |
                       +---------+------+-----+-----
0621      5.0   U2     |P2       |P4           |
                       +---------+--+-----+--------
9999    940.0   I1     |C1
                       +-----------------------#####T
9999  2,437.3   I2     |C2
                       +------------------------######T
9999      0.0   I3     |C3
                       +-----------------------------
9999      0.0   I4     |C4
                       +-----------------------------
TOTAL:  3,397.
   INCR 1.      JUNE    1  4  7 10 13 16 19 22 25 28
             BOTTOM 20% :                    I1I2
                RUN OUT:                     I2I1
COMMAND: change
```

After hitting enter, the schedule is adjusted for the new run lengths as shown in Figure 8.

```
              ABOVE TOP:
                TOP 80% :       I2         I1        I3
AVAIL    COST   UNIT    1  4  7 10 13 16 19 22 25 28
                        .  .  .  .  .  .  .  .  .  .
0625     15.0   U1     |P1       |P3         |P1
                       +---------+-----------+------
0623      5.0   U2     |P2       |P4         |P2
                       +---------+-----------+------
9999      0.0   I1     |C1
                       +-----------------------------
9999    847.0   I2     |C2
                       +--------------------------##T
9999      0.0   I3     |C3
                       +-----------------------------
9999      0.0   I4     |C4
                       +-----------------------------
TOTAL:    867.
   INCR 1.      JUNE    1  4  7 10 13 16 19 22 25 28
             BOTTOM 20% :           I3 I4       I2 I1
                RUN OUT:                           I2
COMMAND: change
```

After hitting enter, the new schedule is displayed. See Figure 10.

```
          ABOVE TOP:
             TOP 80% :      I2    I1      I3
   AVAIL  COST  UNIT   1  4  7  10 13 16 19 22 25 28
                       .  .  .  .  .  .  .  .  .  .
   9999   16.0   U1    |P1         |P3        |P1
                       +-----------+----------+-----
   9999    6.0   U2    |P2         |P4        |P2
                       +-----------+----------+-------
   9999    0.0   I1    |C1
                       +--------------------------
   9999    0.0   I2    |C2
                       +--------------------------
   9999    0.0   I3    |C3
                       +--------------------------
   9999    0.0   I4    |C4
                       +--------------------------
   TOTAL:   22.        .  .  .  .  .  .  .  .  .  .
       INCR 1     JUNE  1  4  7  10 13 16 19 22 25 28
             BOTTOM 20% :        I3 I4
                 RUN OUT:
   COMMAND:change
```

Run outs have been totally avoided. The scheduling costs are the costs for switching from one process to another.

The question still is, however: Is this a cost-minimal solution? When you enter ZAP and hit enter, ZAP comes back with the message "Search completed". Trying TAP results in a similar message: "No improvement found".

This assures us that we found a very good solution.

EXPERIENCES

The current system has been in general use for about one year, and applications have been both in refining and chemicals processing. Refinery applications have focused on the areas of crude scheduling, middle distillate rundown scheduling and blending, lube oil manufacturing and blending, and even hydrofiner scheduling. Examples of chemical plant applications are plastics plants, resins plants, vistalon plants, and solvents plants.

In all cases, manual scheduling with the main scheduling screen has caught on very quickly. There is a high level of enthusiasm among users for the convenience of scheduling in this manner and the impact it has on schedule development time. Schedulers are freed to consider far more alternatives than they could in the past, and claim significant reduction in schedule development time.

There has been some resistance to using the algorithms, however, and it is felt that this is due to a number of reasons. First, they are not well understood by the users. Objectives and response to various cost structures are not clear, and therefore results are sometimes unexpected. This can probably be alleviated through improved documentation and training programs. Another possibility is that it is sometimes difficult to prepare a complete set of cost data. This again can lead to unexpected results, since the algorithms are missing information possessed by the scheduler in the form of his experience.

Even though the algorithms are organized so that the scheduler can still maintain some control over schedule development, it appears that even more may be required. The ability to focus any of the algorithms on a specific unit, for example, or the ability to use different objectives, would be desirable. A final difficulty, in some cases, has been the computer cost for running CPU intensive algorithms. With indiscriminate use, it is easy for a user to spend several thousand dollars in one 30 minute scheduling session. Again, improved understanding and narrower focus would help address this problem.

In conclusion, due to initial positive response and optimistic outlook, an aggressive support and development program is underway. It is felt that providing a tool of this type is feasible, and that significant gains in plant profitability can be achieved though its use.

LITERATURE CITED

1. Baker, T. E., Math Programming, 15, 43 (1981).

2. Godin, V. B., AIIE Transactions, September, 331 (1978).

3. Jones, C. V., P. Robert and K. D. Wurl, "Scheduling Multi-Product Processing Facilities", Master's Thesis, School of Operations Research and Industrial Engineering, Cornell University, Ithica (1980).

APPENDIX

Optional Commands

The following commands allow the user

to invoke optional capabilities.

BLEND. The BLEND command is used if any facilities were defined in STOR that contain mixtures. Here, the initial volumetric composition must be given for each. This table's columns and rows are only those capacities and grades, respectively, that are used for mixtures.

BULK. If supplies or demands occur in bulk deliveries or liftings, as opposed to continuously, they may be described with the BULK command. Here, columns are grades and rows are dates.

CSTOR. If a dummy combined storage capacity was indicated in the STOR screen, the facilities which it includes must be indicated in the CSTOR screen.

DICT. Further description of the codes used in the DEFINE screen can be provided with the DICT command. Eight character mnemonics and descriptors of up to 16 characters may be entered. These descriptions can be displayed on any of the data screens.

DOWN. The DOWN command can be used to include unit shutdown costs if appropriate. Costs are in currency per unit time.

LINCOM. If variable feed rate operation recipes are necessary, the yields may be calculated as a linear combination of the feeds. To describe the equation, the LINCOM command is used. This makes changing feed rates much more convenient, since PROSIT will then automatically update the yields.

OFFSPEC/SWLOSS. The OFFSPEC command allows users to specify volumes of off-spec production which occur when switching from one process to another. A similar command, SWLOSS, can also be used to define time losses due to switching. These losses are deleted from production by the simulator at switching time.

OPCOST. Unit specific production costs for a given operation may be indicated in the OPCOST Table. These data are added to the operation costs (independent of unit) in the OPER Table.

OSCALE. If an operation must be run at different rates on different units, the OSCALE command can be used. Here scaling factors are entered at the intersection of operations and units.

STPR. When a grade can be allocated to more than one storage facility, the STPR command may be used to indicate preferences. In this command the user indicates in which order the facilities are to be filled. Emptying depends on the inventory rule specified on the SYSTEM screen (options are "last in, first out" or "first in, first out").

SWCOST. The cost of switching from one operation to another is specified in the SWCOST Table. Columns are "To" operations while rows are "From" operations, and costs are supplied at appropriate intersections.

TARGET. Inventory targets for grades in demand days or material units may be entered with the TARGET command.

USCALE. If unit shutdowns are planned or if unit efficiencies vary with season, all operations on a given unit for a given date can be scaled (up or down) with the USCALE command.

YIELD. Finally, if feed composition dependent yields were defined in the OPER Table, the YIELD command is required. With this command, the user supplies the volumetric fraction of each product grade which is obtained from one unit of each feed grade. The system uses a linear combination of volumetric fractions to calculate the yields.

MINERALS

- 15 Mineral Engineering Techniques
- 20 Liquid Metals Technology—Part I
- 43 Recent Advances in Ferrous Metallurgy
- 85 Fossil Hydrocarbon and Mineral Processing
- 173 Fundamental Aspects of Hydrometallurgical Processes
- 180 Spinning Wire from Molten Metals

PETROCHEMICALS

- 34 Petrochemicals and Petroleum Refining
- 49 Polymer Processing
- 127 Declining Domestic Reserves—Effect on Petroleum and Petrochemical Industry
- 135 The Petroleum/Petrochemical Industry and the Ecological Challenge
- 142 Optimum Use of World Petroleum
- 212 Interfacial Phenomena In Enhanced Oil Recovery

PETROLEUM PROCESSING

- 34 Petrochemicals and Petroleum Refining
- 54 Hydrocarbons from Oil Shale, Oil Sands, and Coal
- 98 Methanol Technology and Economics
- 103 C_4 Hydrocarbon Production and Distribution
- 127 Declining Domestic Reserves—Effect on Petroleum and Petrochemical Industry
- 135 The Petroleum/Petrochemical Industry and the Ecological Challenge
- 142 Optimum Use of World Petroleum
- 155 Oil Shale and Tar Sands

PHASE EQUILIBRIA

- 2 Pittsburgh and Houston
- 3 Minneapolis and Columbus
- 6 Collected Research Papers
- 81 Phase Equilibria and Related Properties
- 88 Phase Equilibria and Gas Mixtures Properties

PROCESS DYNAMICS

- 36 Process Dynamics and Control
- 46 Process Systems Engineering
- 55 Process Control and Applied Mathematics
- 159 Chemical Process Control
- 214 Selected Topics on Computer-Aided Process Design and Analysis

SEPARATION

- 91 Unusual Methods of Separation
- 120 Recent Advances in Separation Techniques
- 192 Recent Advances in Separation Techniques—II

SONICS

- 1 Ultrasonics—Two Symposia
- 109 Sonochemical Engineering

THERMODYNAMICS

- 7 Applied Thermodynamics
- 44 Thermodynamics
- 140 Thermodynamics—Data and Correlations

TRANSPORT PROPERTIES

- 16 Mass Transfer
- 56 Selected Topics in Transport Phenomena
- 77 Fundamental Research on Heat and Mass Transfer

MISCELLANEOUS

- 26 Chemical Engineering Education—Academic and Industrial
- 48 Chemical Engineering Reviews
- 70 Small-Scale Equipment for Chemical Engineering Laboratories
- 76 High Pressure Technology
- 112 Engineering, Chemistry, and Use of Plasma Reactors
- 125 Vacuum Technology at Low Temperatures
- 143 Standardization of Catalyst Test Methods
- 160 Continuous Polymerization Reactors
- 182 Biorheology
- 183 The Modern Undergraduate Laboratory Innovative Techniques
- 185 Electro organic Synthesis Technology
- 186 Plasma Chemical Processing
- 187 Chronic Replacement of Kidney Function
- 194 Hazardous Chemical—Spills and Waterborne Transportation
- 203 A Review of AIChE's Design Institute for Physical Property Data (DIPPR) and Worldwide Affiliated Activities
- 204 Tutorial Lectures in Electrochemical Engineering and Technology
- 206 Controlled Release Systems

MONOGRAPH SERIES

- 1 Reaction Kinetics by Olaf Hougen
- 2 Atomization and Spray Drying by W. R. Marshall, Jr.
- 3 The Manufacture of Nitric Acid by the Oxidation of Ammonia—The DuPont Pressure Process by Thomas H. Chilton
- 4 Experiences and Experiments with Process Dynamics by Joel O. Hougen
- 5 Present, Past, and Future Property Estimation Techniques by Robert C. Reid
- 6 Catalysts and Reactors by James Wei
- 7 The 'Calculated' Loss-of-Coolant Accident by L. J. Ybarrondo, C. W. Solbrig, H. S. Isbin
- 8 Understanding and Conceiving Chemical Process by C. Judson King
- 9 Ecosystem Technology:Theory and Practice by Aaron J. Teller
- 10 Fundamentals of Fire and Explosion by Daniel R. Stull
- 11 Lumps, Models and Kinetics in Practice by Vern W. Weekman, Jr.
- 12 Lectures in Atmospheric Chemistry by John H. Seinfeld
- 13 Advanced Process Engineering by James R. Fair
- 14 Synfuels from Coal by Bernard S. Lee